Macmillan/McGraw-Hill • Glencoe

Grade 5

Math
Triumphs

Book 3: Geometry, Measurement, and Algebra

Authors

Basich Whitney • Brown • Dawson • Gonsalves • Silbey • Vielhaber

 **Macmillan/McGraw-Hill
Glencoe**

Photo Credits

Cover, i Carl Schneider/Taxi/Getty Images; **iv** (tl)File Photo, (tc tr)The McGraw-Hill Companies, (cl c)Doug Martin, (cr)Aaron Haupt, (bl bc)File Photo; **v** (L to R 1 2 3 4 6 7 8 9 11 12)The McGraw-Hill Companies, (5 10 13 14)File Photo; **vi** Glen Allison/Getty Images; **vii** Donovan Reese/Getty Images; **viii** Scenics of America/PhotoLink/Getty Images; **333** Wes Thompson/CORBIS; **338** Randy Faris/CORBIS; **339** Photodisc/Getty Images; **340** Digital Vision/Alamy; **345** (t)Photodisc/Getty Images, (b)CORBIS; **347** Iconotec/Alamy; **351** Alvis Upitis/Photographer's Choice RF/Getty Images; **355** Cristian Baitg Schreiweis/Alamy; **357** D. Hurst/Alamy; **366** Mike Brinson/Getty Images; **372** Gary Cralle/Getty Images; **378** Ryan McVay/Getty Images; **384** Comstock/CORBIS; **386** Sean Way/Design Pics/CORBIS; **394** Jerry Irwin/Photo Researchers Inc.; **404** John Glustina/Getty Images; **405** Bob Coyle/The McGraw-Hill Companies; **411** IT Stock/PunchStock; **417** Mark Ransom/RansomStudios; **418** Artiga Photo/Masterfile; **425** Radius Images/Jupiter Images; **426** Dinodia Photo Library/Brand X/CORBIS; **433** Image Source Pink/Getty Images; **439** Sandra Ivany/Brand X Pictures/Getty Images; **440** Ken Cavanagh/The McGraw-Hill Companies; **447** Stockdisc/Punchstock

The McGraw·Hill Companies

Macmillan/McGraw-Hill
Glencoe

Send all inquiries to:
Glencoe/McGraw-Hill
8787 Orion Place
Columbus, OH 43240-4027

ISBN: 978-0-07-888206-7
MHID: 0-07-888206-0

Printed in the United States of America.

8 9 10 PRS 16 15 14 13

Math Triumphs
Grade 5, Book 3

Math Triumphs

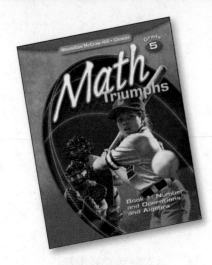

Book 1

Book 2

Book 3

Authors and Consultants

AUTHORS

Frances Basich Whitney
Project Director, Mathematics K–12
Santa Cruz County Office of Education
Capitola, California

Kathleen M. Brown
Math Curriculum Staff Developer
Washington Middle School
Long Beach, California

Dixie Dawson
Math Curriculum Leader
Long Beach Unified
Long Beach, California

Philip Gonsalves
Mathematics Coordinator
Alameda County Office of Education
Hayward, California

Robyn Silbey
Math Specialist
Montgomery County Public Schools
Gaithersburg, Maryland

Kathy Vielhaber
Mathematics Consultant
St. Louis, Missouri

CONTRIBUTING AUTHORS

Viken Hovsepian
Professor of Mathematics
Rio Hondo College
Whittier, California

FOLDABLES Study Organizer **Dinah Zike**
Educational Consultant,
Dinah-Might Activities, Inc.
San Antonio, Texas

CONSULTANTS

Assessment

Donna M. Kopenski, Ed.D.
Math Coordinator K–5
City Heights Educational Collaborative
San Diego, California

Instructional Planning and Support

Beatrice Luchin
Mathematics Consultant
League City, Texas

ELL Support and Vocabulary

ReLeah Cossett Lent
Author/Educational Consultant
Alford, Florida

Reviewers

Each person below reviewed at least two chapters of the Student Edition, providing feedback and suggestions for improving the effectiveness of the mathematics instruction.

Dana M. Addis
Teacher Leader
Dearborn Public Schools
Dearborn, MI

Renee M. Blanchard
Elementary Math Facilitator
Erie School District
Erie, PA

Jeanette Collins Cantrell
5th and 6th Grade Math Teacher
W.R. Castle Memorial Elementary
Wittensville, KY

Helen L. Cheek
K-5 Mathematics Specialist
Durham Public Schools
Durham, NCI

Mercy Cosper
1st Grade Teacher
Pershing Park Elementary
Killeen, TX

Bonnie H. Ennis
Mathematics Coordinator
Wicomico County Public Schools
Salisbury, MD

Sheila A. Evans
Instructional Support Teacher – Math
Glenmount Elementary/Middle School
Baltimore, MD

Lisa B. Golub
Curriculum Resource Teacher
Millennia Elementary
Orlando, FL

Donna Hagan
Program Specialist – Special Programs
    Department
Weatherford ISD
Weatherford, TX

Russell Hinson
Teacher
Belleview Elementary
Rock Hill, SC

Tania Shepherd Holbrook
Teacher
Central Elementary School
Paintsville, KY

Stephanie J. Howard
3rd Grade Teacher
Preston Smith Elementary
Lubbock, TX

Rhonda T. Inskeep
Math Support Teacher
Stevens Forest Elementary School
Columbia, MD

Albert Gregory Knights
Teacher/4th Grade/Math Lead Teacher
Cornelius Elementary
Houston, TX

Barbara Langley
Math/Science Coach
Poinciana Elementary School
Kissimmee, FL

David Ennis McBroom
Math/Science Facilitator
John Motley Morehead Elementary
Charlotte, NC

Jan Mercer, MA; NBCT
K-5 Math Lab Facilitator
Meadow Woods Elementary
Orlando, FL

Rosalind R. Mohamed
Instructional Support Teacher – Mathematics
Furley Elementary School
Baltimore, MD

Patricia Penafiel
Teacher
Phyllis Miller Elementary
Miami, FL

Lindsey R. Petlak
2nd Grade Instructor
Prairieview Elementary School
Hainesville, IL

Lana A. Prichard
District Math Resource Teacher K-8
Lawrence Co. School District
Louisa, KY

Stacy L. Riggle
3rd Grade Spanish Magnet Teacher
Phillips Elementary
Pittsburgh, PA

Wendy Scheleur
5th Grade Teacher
Piney Orchard Elementary
Odenton, MD

Stacey L. Shapiro
Teacher
Zilker Elementary
Austin, TX

Kim Wilkerson Smith
4th Grade Teacher
Casey Elementary School
Austin, TXL

Wyolonda M. Smith, NBCT
4th Grade Teacher
Pilot Elementary School
Greensboro, NC

Kristen M. Stone
3rd Grade Teacher
Tanglewood Elementary
Lumberton, NC

Jamie M. Williams
Math Specialist
New York Mills Union Free School District
New York Mills, NY

Contents

Chapter 8 **Geometry**

Badlands National Park, Interior, South Dakota

Chapter 9 Area

City Hall, Philadelphia, Pennsylvania

Contents

Surface Area, Volume, and Measurement

The Alamo, San Antonio, Texas

SCAVENGER HUNT

Let's Get Started

Use the Scavenger Hunt below to learn where things are located in each chapter.

1. What is the title of Chapter 9?

2. What is the Key Concept of Lesson 8-4?

3. What is the definition of a parallelogram on page 334?

4. What are the vocabulary words for Lesson 10-2?

5. How many examples are presented in the Chapter 8 Study Guide?

6. What figure is used as a model in Example 1 on page 368?

7. According to Lesson 10-2, how many ounces are in one gallon?

8. What do you think is the purpose of the Progress Check 1 on page 348?

9. On what pages will you find the study guide for Chapter 10?

10. In Chapter 9, find the internet address that tells you where you can take the Online Readiness Quiz.

Geometry

There are triangles, circles, and figures all around us.

You see figures every day. Everywhere you go there are different figures. Some may be flat, or solid. Many windows and buildings form basic figures.

STEP 1 Quiz

Are you ready for Chapter 8? Take the Online Readiness Quiz at *glencoe.com* to find out.

STEP 2 Preview

Get ready for Chapter 8. Review these skills and compare them with what you'll learn in this chapter.

What You Know	What You Will Learn
You know how to describe and recognize some figures. **TRY IT!** Name the figures. 1 ▢ _____ 2 △ _____	*Lessons 8-1, 8-2* Triangles and quadrilaterals can be classified by their angles and sides. A parallelogram has four sides. Each pair of opposite sides is parallel and equal in length. An equilateral triangle has three sides equal in length and three angles that are the same measure.
You know how to identify a circle. ◀ circle	*Lesson 8-3* A **circle** has a **radius** and a **diameter**. The diameter is twice as long as the radius. radius diameter
You know that two-dimensional figures are flat.	*Lesson 8-4* **Three-dimensional figures** are not flat.

Quadrilaterals

KEY Concept

Quadrilaterals have four sides and four angles. Some quadrilaterals have special names.

Type	Example	Description
rectangle		A rectangle has four right angles, with two pairs of equal sides.
square		A square has four right angles. All sides are equal.
parallelogram		The opposite sides of a parallelogram are parallel and equal in length.
rhombus *These marks show equal sides.*		All four sides of a rhombus are equal. Opposite sides are parallel.
trapezoid		A trapezoid has only one pair of opposite sides parallel.

VOCABULARY

parallel lines
    lines that are the same distance apart; parallel lines do not meet or cross

quadrilateral
    a figure that has four sides and four angles

Quadrilaterals can be classified by the size of their angles and the length of their sides.

Example 1

Classify the figure in as many ways as possible.

1. Look at the figure.

2. Are opposite sides equal?
   Yes

3. Are any of the opposite sides parallel?
   Yes

4. Does the figure have exactly one pair of parallel sides?
   No

5. The figure can be classified as a parallelogram, rectangle, square, and rhombus.

YOUR TURN!

Classify the figure in as many ways as possible.

1. Look at the figure.

2. Are opposite sides equal?

   _____

3. Are any of the opposite sides parallel?

   _____

4. Does the figure have exactly one pair of parallel sides?

   _____

5. The figure can be classified as a

   _____.

Example 2

Identify the figure.

1. The figure has four sides.
   The figure is a quadrilateral.

2. All sides are equal in length.
   The figure is a rhombus or a square.

3. All angles are right angles.
   The figure is a square.

YOUR TURN!

Identify the figure.

1. The figure has _____ sides.

   The figure is a(n) _____.

2. There is _____ pair of parallel sides.

3. There are _____ right angles.

   The figure is a(n) _____.

Who is Correct?

Draw a parallelogram.

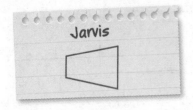

Circle correct answer(s). Cross out incorrect answer(s).

▶ Guided Practice

Classify each quadrilateral in as many ways as possible.

1

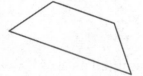

  1. Are opposite sides equal? _____

  2. Are any of the opposite sides parallel? _____

  3. Does the figure have exactly one pair
      of parallel sides? _____

  4. The shape is a _____

  _____.

2

  1. Are opposite sides equal? _____

  2. Are any of the opposite sides parallel? _____

  3. Does the figure have exactly one pair of parallel
      sides? _____

  4. The shape is a _____

  _____.

Step by Step Practice

3 Identify the figure.

Step 1 The figure has _____ sides.

The figure is a(n) _____.

Step 2 Are any of the sides parallel? _____

Step 3 There are _____ pairs of parallel sides.

The figure is a(n) _____.

Identify the figure.

4 The figure has _____ sides.

The figure is a(n) _____.

Are any of the sides parallel and equal? _____

There are _____ pairs of equal and parallel sides.

The figure is a(n) _____.

5 The figure has _____ sides.

The figure is a(n) _____.

There are _____ pairs of equal and parallel sides.

The figure is a(n) _____.

Step by Step Problem-Solving Practice

Solve.

6 **TILES** Pete is placing new tile on his kitchen floor. What quadrilateral figure describes the dark blue part of the tile Pete is using?

Understand Read the problem. Write what you know. The figure of the pattern on the tile is a _____.

Plan Pick a strategy. One strategy is to use a diagram.

Solve Trace the outline of the figure. There are _____ sides. There are _____ pairs of parallel sides. The figure is a(n) _____.

Check Review the definition of the figure you named.

7 **SKETCHING** Anya is sketching a figure in art class. Anya's figure has four sides and four right angles. There are two sets of lines equal in length. Opposite sides are parallel. What is the name of the figure Anya is sketching? Check off each step.

_____ Understand: I circled key words.

_____ Plan: To solve the problem, I will _____.

_____ Solve: The answer is a(n) _____.

_____ Check: I checked my answer by _____.

8 **Reflect** The word *quadrilateral* has two parts. *Quad* means "four" and *lateral* means "side or relating to the side." In your own words, explain the meaning of the word quadrilateral.

_____

_____

9 Circle the quadrilaterals.

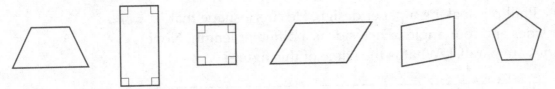

10 Circle the rectangles.

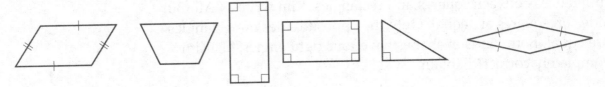

11 Circle the parallelograms.

Identify each figure.

12

_____

13

_____

14

ROAD
CLOSED

_____

15

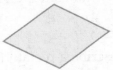

_____

GO ON

Identify each figure.

16

_____

17

_____

18 **DESIGN**   Phillip is making a paper design. He cuts a shape that has four sides and four angles. Each side is a different length. None of the sides are parallel. What is the name of this figure?

_____

19 **HOMEWORK**   Kim's math homework says to explain the differences between squares and rectangles. Kim writes, "All four sides on squares are equal. Only the opposite sides on rectangles are equal. Both kinds of shapes have four right angles." Is Kim completely correct? Explain.

_____

_____

Vocabulary Check   **Write the vocabulary word that completes each sentence.**

20 A(n) _____ has four sides and four angles.

21 _____ are lines that are the same distance apart; parallel lines do not meet or cross.

22 A rectangle with four equal sides and four equal angles is a(n) _____ .

23 **Writing in Math**   Describe a rectangle in words.

_____

_____

_____

STOP

Triangles

KEY Concept

Types of triangles are classified by the measures of their angles or the lengths of their sides.

right triangle
one 90° angle

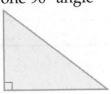

acute triangle
three angles less than 90°

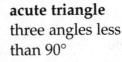

obtuse triangle
one angle greater than 90°

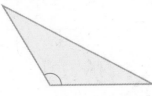

isosceles triangle
at least two sides are congruent

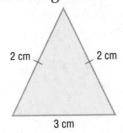

2 cm   2 cm

3 cm

equilateral triangle
all sides are congruent

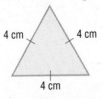

4 cm   4 cm

4 cm

scalene triangle
no congruent sides

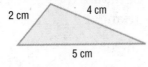

2 cm   4 cm

5 cm

VOCABULARY

acute angle
an angle with a measure greater than 0° and less than 90°

congruent
having the same size, shape, or measure

obtuse angle
an angle that measures greater than 90° but less than 180°

Triangles can be named by their angles and sides.
For example, a triangle with one right angle and
two congruent sides is called an isosceles right triangle.

Example 1

Name the triangle by its sides.

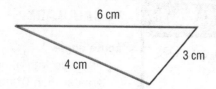

1. The triangle has **no** congruent sides.
2. The triangle is a **scalene triangle**.

Name the triangle by its sides.

1. The triangle has _____ congruent sides.
2. The triangle is a(n) _____.

Example 2

Name the triangle by its angles.

1. Use a right angle to show 90°. Compare the right angle to the angles of the triangle.
2. There are **three** angles less than 90°.
3. There are **no** angles greater than 90°.
4. There are **no** right angles.
5. The triangle is an **acute triangle**.

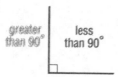

Name the triangle by its angles.

1. Use a right angle to show 90°. Compare the right angle to the angles of the triangle.
2. There are _____ angles less than 90°.
3. There is _____ angle greater than 90°.
4. There are _____ right angles.
5. The triangle is a(n) _____ triangle.

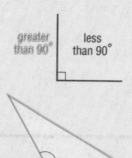

Who is Correct?

Draw a right triangle.

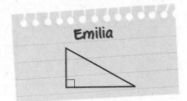

Emilia

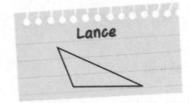

Lance

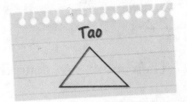

Tao

Circle correct answer(s). Cross out incorrect answer(s).

 Guided Practice

Name each triangle by its sides.

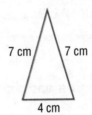

7 cm   7 cm

4 cm

1 The triangle has _____ congruent sides.

The triangle is a(n) _____.

2

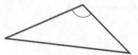

5 in.   5 in.

5 in.

_____

3

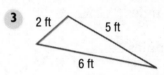

2 ft   5 ft

6 ft

_____

Name the triangle by its angles.

4 There are _____ angles less than 90°.

There are _____ angles greater than 90°.

There is _____ right angle.

The triangle is a(n) _____.

5

_____

6

_____

GO ON

7 Name the triangle by its angles and sides.

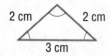

2 cm   2 cm
3 cm

Step 1   There are _____ congruent sides.

Step 2   There are _____ acute angles.

Step 3   There are _____ right angles.

Step 4   There are _____ obtuse angles.

Step 5   The triangle is a(n) _____ triangle.

Name each triangle by its angles and sides.

8   number of congruent sides _____

number of acute angles _____

number of right angles _____

number of obtuse angles _____

The triangle is a(n) _____ triangle.

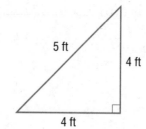

5 ft   4 ft
4 ft

9

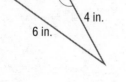

3 in.
4 in.
6 in.

_____

10

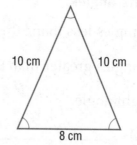

10 cm   10 cm
8 cm

_____

Step by Step Problem-Solving Practice

Solve.

11 SAFETY Reginald saw a street sign that looked like the one at the right. Classify the shape of the sign.

Understand Read the problem. Write what you know.

Reginald saw _____.

Plan Pick a strategy. One strategy is to use a diagram.

Solve The street sign has _____ sides and

_____ angles.

The figure is a(n) _____.

The sides appear to be _____ in length.

The figure is a(n) _____.

Check Review the definition of the figure you named.

12 MATH Madison's math teacher drew a triangle on the board and asked her to name the shape. Madison saw that each side of the triangle was a different length. What kind of triangle did the math teacher draw? Check off each step.

_____ Understand: I circled key words.

_____ Plan: To solve the problem, I will _____.

_____ Solve: The answer is _____.

_____ Check: I checked my answer by _____.

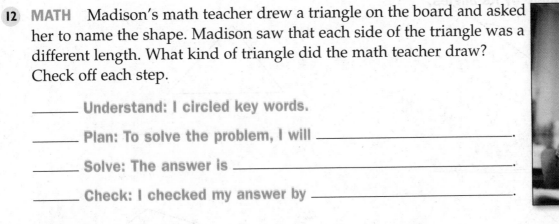

13 Reflect Can a triangle be both isosceles and obtuse? Explain.

_____

_____

Skills, Concepts, and Problem Solving

Name each triangle by its sides.

14
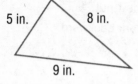
5 in. 8 in.
9 in.

_____

15

7 ft 5 ft
5 ft

_____

16
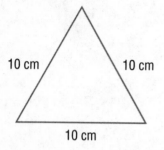
10 cm 10 cm
10 cm

_____

17
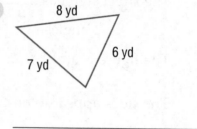
8 yd
7 yd 6 yd

_____

Name each triangle by its angles.

18
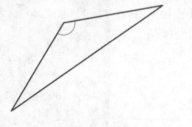

_____

19
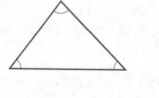

_____

20
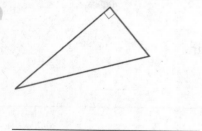

_____

21
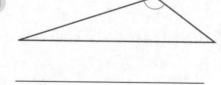

_____

Name the triangle by its angles and sides.

22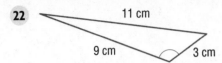

11 cm

9 cm   3 cm

_____

23 **DRAWING**   Violeta draws an equilateral triangle with two sides that each measure 6 inches. What is the length of the third side?

_____

24 **LUNCH**   Peyton shares a sandwich with his sister for lunch. He cuts the sandwich into two triangles. Look at the picture at the right. What type of triangles is the sandwich cut into?

_____

Vocabulary Check   **Write the vocabulary word that completes each sentence.**

25 A(n) _____ has three equal sides.

26 _____ means having the same size, shape, or measure.

27 A triangle with a right angle is a(n) _____.

28 **Writing in Math**   Can a triangle be both a right triangle and an acute triangle?

_____

_____

_____

 Spiral Review

Identify each figure. (Lesson 8-1, p 334)

29

30

_____   _____

STOP

Write the name of each figure.

1

_____

2

_____

3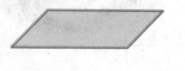

_____

4

_____

Name each triangle by its sides or angles.

5

3 cm   3 cm

3 cm

_____

6

2 in.   5 in.

7 in.

_____

7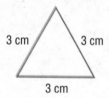

_____

8

_____

Solve.

9 **FIGURES** Jamie is drawing a quadrilateral figure. The top and bottom are not equal in length. There is one pair of parallel sides. What name would you give to the figure Jamie draws?

_____

10 Can a right triangle also be equilateral? Explain.

_____

_____

Circles

KEY Concept

A **circle** is the set of all points in a plane that are the same distance from a point called the center.

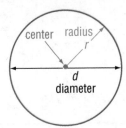

The letter *d* is used to represent the **diameter**.
The letter *r* is used to represent the **radius**.

VOCABULARY

circle
a closed figure in which all points are the same distance from a fixed point called the center

diameter
the distance across a circle through its center

radius
the distance from the center to any point on a circle

A diameter is 2 times as long as the radius.
The radius is $\frac{1}{2}$ as long as the diameter.

Example 1

Find the radius of the circle.

1. The radius is $\frac{1}{2}$ as long as the diameter.

2. One-half of 12 cm is 6 cm.

   12 cm ÷ 2 = 6 cm

3. The radius of the circle is 6 cm.

YOUR TURN!

Find the radius of the circle.

1. The radius is _____ as long as the diameter.

2. One-half of _____ is _____.

   _____ ÷ _____ = _____

3. The radius of the circle is _____.

GO ON

Who is Correct?

A circle has a diameter of 6 inches. What is the radius of the circle?

Elvio
6 in. × 2 = 12 in.

Carisa
r = d
r = 6 inches

Jenna
6 in. ÷ 2 = 3 in.
r = 3 in.

Circle correct answer(s). Cross out incorrect answer(s).

 Guided Practice

Find the radius of each circle.

1 _____ ÷ _____ = _____

$d = 10$ in. $r =$ _____

2 _____ ÷ _____ = _____

$d = 4$ cm $r =$ _____

Step by Step Practice

3 Find the diameter of the circle.

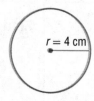

$r = 4$ cm

Step 1 The diameter is _____ times as long as the radius.

Step 2 Two times _____ is _____.

_____ × _____ = _____

Step 3 The diameter of the circle is _____.

Find the diameter of each circle.

4 _____ × _____ = _____

$r = 7$ cm $d =$ _____

5 _____ × _____ = _____

$r = 9$ in. $d =$ _____

Step by Step Problem-Solving Practice

Solve

Problem-Solving Strategies
☐ Use a diagram.
☐ Look for a pattern.
☐ Guess and check.
☐ Solve a simpler problem.
☑ Work backward.

6 PLAYGROUND Tionna wants to find the radius of the center circle on a basketball court. The line through the center of the circle measures 12 feet. What is the radius?

Understand Read the problem. Write what you know.

Tionna wants to find the _____ of a circle.

Plan Pick a strategy. One strategy is to work backward.

Solve The diameter is _____ times the radius.

The radius is _____ the diameter.

_____ ÷ _____ = _____

The radius of the circle is _____.

Check Use multiplication to check.

7 PAINTING Hassan is painting a large circle with a radius of 3 ft. What is the diameter of Hassan's painting? Check off each step.

_____ Understand: I circled key words.

_____ Plan: To solve the problem, I will _____.

_____ Solve: The answer is _____.

_____ Check: I checked my answer by _____.

8 Reflect Explain the difference between the radius and the diameter of a circle.

_____

_____

▶ Skills, Concepts, and Problem Solving

Find the radius and diameter of each circle.

9
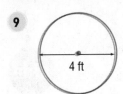
4 ft

radius: _____ ft
diameter: _____ ft

10
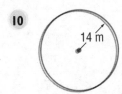
14 m

radius: _____ m
diameter: _____ m

GO ON

Find the radius or diameter.

11 $d = 10$ cm

  $r = $ _____

12 $d = 22$ yd

  $r = $ _____

13 $r = 8$ cm

  $d = $ _____

14 $r = 4$ in

  $d = $ _____

15 $r = 3$ yd

  $d = $ _____

16 $d = 40$ ft

  $r = $ _____

17 **HOBBIES** Dario collects circular clocks. The face of his largest clock has a radius of 23 inches. What is the diameter of the clock?

_____

18 **SIZE** Circle A has a radius of 10 cm. Circle B has a diameter of 24 cm. Which circle is larger? _____

Vocabulary Check **Write the vocabulary word that completes each sentence.**

19 The distance across a circle through its center is called the

  _____.

20 **Writing in Math** Name some objects that have a diameter or a radius.

_____

Spiral Review

Solve. (Lesson 8-1, p. 334)

21 **DRAWING** Eva draws a quadrilateral. All four sides have the same length. Its opposite sides are parallel. What quadrilateral did she draw?

_____

Name each triangle by its angles. (Lesson 8-2, p. 341)

22

_____

23

_____

STOP

Three-Dimensional Figures

KEY Concept

Three-dimensional figures are named by the types of surfaces they have. Their surfaces can be curved, flat, or both.

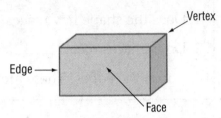

The base is an "end" face. The bases are shaded below.

Figure	Example	Description
rectangular prism		a prism with six rectangular faces
cube		a prism with six faces that are congruent squares
triangular prism		a prism that has triangular bases
cone		a solid that has a circular base and one curved surface from the base to a vertex
cylinder		a solid with two parallel, congruent, circular bases; a curved surface connects the bases
sphere		a solid figure that is a set of all points that are the same distance from the center

VOCABULARY

edge
the line segment where two faces of a three-dimensional figure meet

face
the flat part of a three-dimensional figure

three-dimensional figure
figure that has length, width, and height

vertex
the point on a three-dimensional figure where three or more edges meet

GO ON

Example 1

Find the number of edges.

1. Does the shape have faces? yes

2. Do the faces meet? yes

3. Count the edges. There are 12 edges.

YOUR TURN!

Find the number of edges.

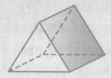

1. Does the shape have faces? _____

2. Do the faces meet? _____

3. Count the edges. There are _____ edges.

Example 2

Identify the three-dimensional figure.

1. Is the figure flat, curved, or both? both

2. Describe the base(s). one circular base

3. The figure is a cone.

YOUR TURN!

Identify the three-dimensional figure.

1. Is the figure flat, curved, or both? _____

2. Describe the base(s).

_____

3. The figure is a(n) _____.

Who is Correct?

Identify the three-dimensional figure.

Debbie
cone

Ben
rectangle

Tia
cylinder

Circle correct answer(s). Cross out incorrect answer(s).

 Guided Practice

Find the number of edges.

1. Count the edges. There are _____ edges.

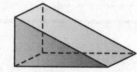

2) Identify the three-dimensional figure.

Step 1 Is the figure flat, curved, or both? _____

Step 2 Describe the base(s). _____

Step 3 The figure is a(n) _____.

Identify each three-dimensional figure.

3)

The figure is _____.

Describe the base(s). _____

The figure is a(n) _____.

4)

The figure is _____.

Describe the base(s). _____

The figure is a(n) _____.

Step (by) **Step Problem-Solving Practice**

Solve.

5) **MUSEUM** Isabel was looking at a structure at a museum. The structure was a large three-dimensional figure. What is the name of the figure?

Problem-Solving Strategies
☑ Use a diagram.
☐ Look for a pattern.
☐ Guess and check.
☐ Solve a simpler problem.
☐ Work backward.

Understand Read the problem. Write what you know. Isabel was looking at a _____.

Plan Pick a strategy. One strategy is to use a diagram.

Solve Look at the photo. Describe the shape.

_____.

_____.

The three-dimensional figure is a _____.

Check Compare the figure to other figures in the lesson.

GO ON

6 **ART** Rose's art teacher set a soup can on the table. She asked students to draw the shape of the can. What figure did Rose draw? Check off each step.

_____ Understand: I circled key words.

_____ Plan: To solve the problem, I will _____.

_____ Solve: The answer is _____.

_____ Check: I will check my answer by _____.

7 **DINNER** Benito's family had spaghetti and meatballs for dinner. He wanted to identify the shape of the meatballs. What figure are the meatballs?

_____

8 **Reflect** What is the difference between three-dimensional figures and two-dimensional figures? Explain. Give an example of each.

_____

_____

_____

▶ Skills, Concepts, and Problem Solving

Identify each three-dimensional figure.

9

10

11

_____

12

13

14
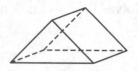

_____   _____   _____

Read each description. Identify the three-dimensional figure.

15 This figure has a curved surface, a circular top, and a circular shaped base.

_____

16 This figure has six rectangular faces. _____

Vocabulary Check   **Write the vocabulary word that completes each sentence.**

17 A figure that has length, width, and height is a(n)

_____.

18 A(n) _____ is the line segment where two faces of a three-dimensional figure meet.

19 **Writing in Math**   Explain why figures with curved surfaces cannot be prisms.

_____

_____

 Spiral Review

Find the radius or diameter. (Lesson 8-3, p. 349)

20 $r = 8$ cm

$d =$ _____

21 $d = 24$ yd

$r =$ _____

22 $d = 10$ ft

$r =$ _____

Identify each figure. (Lesson 8-1, p. 334)

23

_____

24

_____

PLAYGROUND This playground structure is composed of three-dimensional figures.

Find the radius and diameter of each circle.

1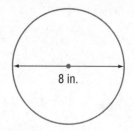

8 in.

radius: _____ in.
diameter: _____ in.

2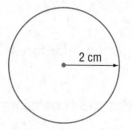

2 cm

radius: _____ cm
diameter: _____ cm

Find the radius or diameter.

3 $d = 12$ cm

$r =$ _____

4 $d = 20$ yd

$r =$ _____

5 $r = 8$ cm

$d =$ _____

6 $r = 11$ in.

$d =$ _____

7 $r = 3$ yd

$d =$ _____

8 $d = 30$ ft

$r =$ _____

Identify each three-dimensional figure.

9

_____

10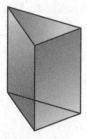

_____

11 BASEBALL According to baseball regulations, the pitcher's mound, which is circular, must have a diameter of 18 feet. What is the radius of a pitcher's mound?

_____

12 Ms. Jackson asked her students to draw three-dimensional figures. She told students to use objects as models for their figures. Maurice decided to draw a cereal box. What three-dimensional figure did Maurice draw?

_____

Vocabulary and Concept Check

acute angle, *p. 341*

circle, *p. 349*

congruent, *p.341*

diameter, *p. 349*

edge, *p. 353*

face, *p. 353*

obtuse angle, *p. 341*

parallel lines, *p.334*

quadrilateral, *p. 334*

radius, *p. 349*

three-dimensional figure, *p. 353*

vertex, *p. 353*

Write the vocabulary word that completes each sentence.

1 A(n) _____ is the line segment where two faces of a three-dimensional figure meet.

2 The distance across a circle through its center is called a(n) _____.

3 _____ means having the same size, shape, or measure.

4 A(n) _____ is a figure that has four sides and four angles.

5 An angle that measures greater than 90° but less than 180° is called a(n) _____.

Label each diagram below. Write the correct vocabulary term in each blank.

6 _____

7 _____

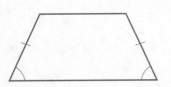

8 _____

Lesson Review

8-1 Quadrilaterals (pp. 334-340)

Identify each figure.

9

_____

10

_____

Identify the figure.

1. The figure has four sides.

   The figure is a(n) quadrilateral.

2. There is one pair of parallel sides.

3. There are no right angles.

   The figure is a(n) trapezoid.

8-2 Triangles (pp. 341-347)

Name each triangle by its sides or angles.

11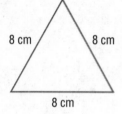

8 cm   8 cm

8 cm

_____

12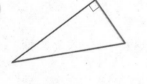

_____

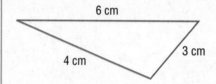

Name the triangle by its sides.

6 cm

4 cm   3 cm

1. The triangle has no congruent sides.

2. The triangle is a scalene triangle.

8-3 Circles (pp. 349-352)

Find the radius or diameter of the circle.

13

radius: _____ in.

14

diameter: _____ yd

Example 3

Find the radius of the circle.

1. The radius is $\frac{1}{2}$ as long as the diameter.

2. One-half of 10 cm is 5 cm.

8-4 Three-Dimensional Figures (pp. 353-357)

Identify each three-dimensional figure.

15

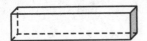

_____

16

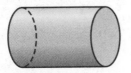

_____

17

_____

Example 4

Identify the three-dimensional figure.

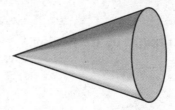

1. Is the shape flat, curved, or both? **both**

2. Describe the base(s). **one circular base**

3. The three-dimensional shape is a cone.

Identify each figure.

1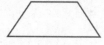

_____

2

_____

3

_____

4

_____

Name each triangle by its sides.

5

4 ft

5 ft

7 ft

_____

6

7 cm   7 cm

7 cm

_____

Name each triangle by its angles.

7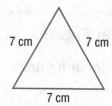

_____

8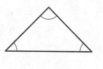

_____

Find the radius or diameter.

9 $d = 14$ cm

   $r =$ _____

10 $d = 22$ yd

   $r =$ _____

11 $r = 3$ cm

   $d =$ _____

12 $r = 15$ in.

   $d =$ _____

13 $r = 8$ yd

   $d =$ _____

14 $d = 8$ ft

   $r =$ _____

GO ON

Identify each three-dimensional figure.

15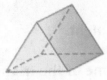

_____

16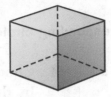

_____

17

_____

18

_____

19

_____

20

_____

Solve.

21 **DRAWING** Mitchell is drawing a triangle. He makes each side of the triangle a different length. What kind of triangle did Mitchell draw?

_____

22 **COMPASS** Naomi is using a compass to draw a circle. When she is finished, Naomi uses a ruler to measure the diameter of her circle. The diameter is 18 inches. What is the radius of Naomi's circle?

_____

Correct the mistakes.

23 Mrs. Reyes asked students in her math class to draw three-dimensional figures. She set some objects on a table. She told students to use the objects as models for their figures. Paquito decided to use the roll of paper towels as a model. Look at Paquito's drawing. What did he do wrong?

_____

_____

STOP

Choose the best answer and fill in the corresponding circle on the sheet at right.

1 Look at these figures. Which figure is a quadrilateral?

A

C

B

D

2 Choose the correct name of the figure.

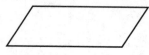

A square

B triangle

C rectangle

D parallelogram

3 Choose the correct name of the triangle by the measure of its sides.

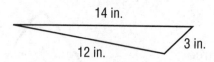

14 in.

12 in.   3 in.

A equilateral triangle

B isosceles triangle

C scalene triangle

D none of the above

4 Choose the correct name of the triangle by the measure of its angles.

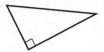

A acute triangle

B right triangle

C obtuse triangle

D none of the above

5 Which of the following shows an isosceles right triangle?

A

5 in.

5 in.

B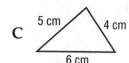

C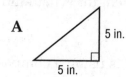

5 cm   4 cm

6 cm

D

6 Choose the correct diameter.

$r = 3$

$d = ?$

A 3   C 9

B 6   D 12

GO ON

7 Choose the correct radius.

$$d = 18$$

$$r = ?$$

A 9   **C** 18

B 27   **D** 36

---

8 Which of the following best describes the shape of a soccer ball?

    A cone

    B rectangular prism

    C sphere

    D vertex

---

9 Choose the correct name of the three-dimensional figure.

    A square   **C** rectangle

    B cube   **D** box

---

10 Which of the following figures shows a cone?

    A   **C**

    B   **D**

11 Which of the following describes the object below?

    A triangular prism

    B sphere

    C cube

    D rectangular prism

1	Ⓐ	Ⓑ	Ⓒ	D
2	Ⓐ	Ⓑ	Ⓒ	Ⓓ
3	Ⓐ	Ⓑ	Ⓒ	Ⓓ
4	Ⓐ	Ⓑ	Ⓒ	Ⓓ
5	Ⓐ	Ⓑ	Ⓒ	Ⓓ
6	Ⓐ	Ⓑ	Ⓒ	Ⓓ
7	Ⓐ	Ⓑ	Ⓒ	Ⓓ
8	Ⓐ	Ⓑ	Ⓒ	Ⓓ
9	Ⓐ	Ⓑ	Ⓒ	Ⓓ
10	Ⓐ	Ⓑ	Ⓒ	Ⓓ
11	Ⓐ	Ⓑ	Ⓒ	Ⓓ

Success Strategy

After answering all the questions, go back and check your work. Make sure you circled the correct answer to each problem.

STOP

Area

Why *is* area important?

What if you compare your volleyball team's record with your cousin's, but the two of you play on different-sized courts? It would be an unfair comparison. Most school club teams play on volleyball courts that are of regulation size.

STEP **1** Quiz

Are you ready for Chapter 9? Take the Online Readiness Quiz at *glencoe.com* to find out.

STEP **2** Preview

Get ready for Chapter 9. Review these skills and compare them with what you'll learn in this chapter.

What You Know	What You Will Learn
You know how to measure the lengths of items. 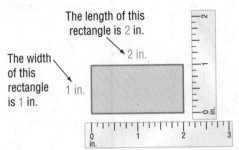	*Lesson 9-2* To find the area of a rectangle, you multiply the length by the width. Area = length × width Area = 2 inches × 1 inch Area = 2 square inches
You know that you can make a parallelogram into a rectangle. Step 1: Step 2: Step 3:	*Lesson 9-3* To find the area of a parallelogram, multiply the base by the height. Area = base × height Area = 3 inches × 2 inches Area = 6 square inches 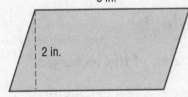
You know you can separate a rectangle into two triangles. 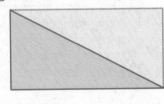	*Lesson 9-4* A triangle is half of a rectangle. So the area of a triangle is one-half the area of a rectangle with the same base and height. Area = base × height Area = ½ × base × height 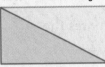

Introduction to Area

KEY Concept

The **area** of a figure is the number of **square units** needed to cover a surface.

To find the area of a figure, you can count the number of square units the figure covers.

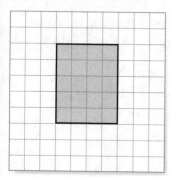 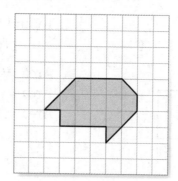

The area of the rectangle is 20 square units.

The area of the figure is about $14\frac{1}{2}$ square units.

VOCABULARY

area
   the number of square units needed to cover a region or plane figure

square unit
   a unit for measuring area

The units of area are square units.

Example 1

Find the area of the rectangle.

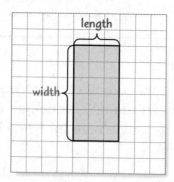

Count the number of squares the rectangle covers.

The area is 18 square units.

YOUR TURN!

Find the area of the rectangle.

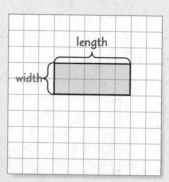

Count the number of squares the rectangle covers. The area is _____ square units.

Example 2

Estimate the area of the figure.

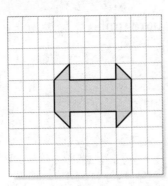

1. Count the number of whole squares the figure covers. **10 whole squares**

2. Count the number of half squares the figure covers. **4 half squares**

    Convert the half squares to whole squares.

    $\frac{1}{2} + \frac{1}{2} + \frac{1}{2} + \frac{1}{2} = 2$

3. Add the number of whole squares.
    10 + 2 = 12

The area of the figure is about 12 square units.

YOUR TURN!

Estimate the area of the figure.

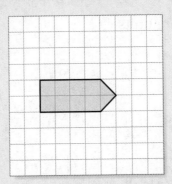

1. Count the number of whole squares the figure covers. _____ whole squares

2. Count the number of half squares the figure covers.

    _____ half square(s) = _____ whole square(s)

3. Add the number of whole squares.

    _____ + _____ = _____

The area of the figure is about

    _____ square units.

Who is Correct?

Find the area of the square.

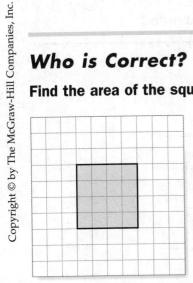

Darcy
4 square units

Alek
8 square units

Una
16 square units

Circle correct answer(s). Cross out incorrect answer(s).

Draw a figure that has the given area.

1 14 square units

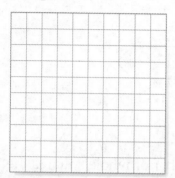

2 25 square units

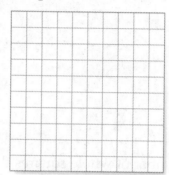

Step by Step Practice

3 Estimate the area of the figure.

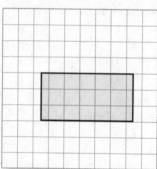

Step 1 Count the number of whole squares.

_____ whole squares

Step 2 Count the number of half squares.

_____ half square(s) = _____ whole square(s)

Step 3 Add the number of whole squares.

_____ + _____ = _____

The area is about _____ square units.

Estimate the area of each figure.

4

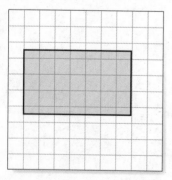

_____ whole square(s)

_____ half square(s)

The area is about

_____.

5

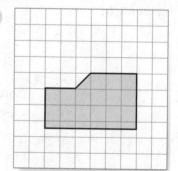

The area is about

_____.

Step by Step Problem-Solving Practice

Solve.

6 **GEOMETRY** What is the area of a rectangle that has sides of 8 units and 6 units?

Understand Read the problem. Write what you know.

A rectangle has sides of _____ units and

_____ units.

Plan Pick a strategy. One strategy is to draw a diagram. Draw a rectangle that has sides of 8 units and 6 units.

Solve Count the number of squares the figure covers.

The area of the rectangle is _____ square units.

Check Add the number of squares in each row.

7 **ROOMS** Talia's dining room floor measures 10 feet by 8 feet. What is the area of the floor? Check off each step.

_____ Understand: I circled key words.

_____ Plan: To solve the problem, I will _____.

_____ Solve: The answer is _____.

_____ Check: I checked my answer by _____.

8 **SHOPPING** Howie bought a blanket. It is 7 meters wide and 8 meters long. What is the area of the blanket?

_____

9 **Reflect** Look at the figure at the right. Is the area 23 square units? Explain.

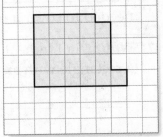

_____

_____

GO ON

▶ Skills, Concepts, and Problem Solving

Draw a figure that has the given area.

10 16 square units

11 30 square units

Find the area of each figure.

12 The area of the figure is about _____.

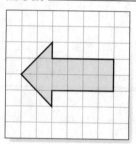

13 The area of the figure is

_____.

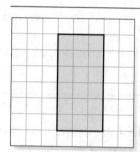

Solve.

14 **ART**   Look at the photo at the right. It shows the largest picture Gregory painted. What is the area?

_____

Vocabulary Check   **Write the vocabulary word that completes each sentence.**

15 A(n) _____ is a unit for measuring area.

16 _____ is the number of square units needed to cover a region or plane figure.

17 **Writing in Math**   Explain how to find the length and width of a rectangle with an area of 24 square units.

ART Gregory's painting is 8 inches by 10 inches.

_____

_____

_____ **STOP**

Area of a Rectangle

KEY Concept

Find the **area** of a **rectangle** using the formula below.

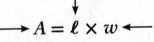

ℓ is the length of
the rectangle.

A is the area of
the rectangle. → $A = \ell \times w$ ← w is the width of the rectangle.

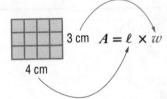

3 cm $A = \ell \times w$

4 cm

The area of the rectangle is 12 square centimeters or 12 cm².

The units of area are **square units**. Remember that a
square is a special rectangle because a square has four
equal sides.

VOCABULARY

area
the number of square
units needed to cover a
region or a plane figure

rectangle
a quadrilateral with four
right angles; opposite
sides are parallel and
equal in length

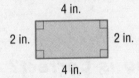

4 in.

2 in.   2 in.

4 in.

square
a rectangle with
equal sides

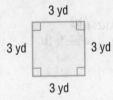

3 yd

3 yd   3 yd

3 yd

square unit
a unit for measuring area

Example 1

Find the area of the rectangle.

1. The length is 5 inches.
   The width is 3 inches.

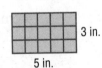

3 in.

5 in.

2. Substitute these values into
   the formula. Multiply.

   $A = \ell \times w$
   $A = 5 \text{ in.} \times 3 \text{ in.}$
   $A = 15 \text{ in}^2$

The area of the rectangle is 15 square inches.

GO ON

YOUR TURN!

Find the area of the rectangle.

1. The length is _____ yards.
   The width is _____ yards.

2. Substitute these values into the formula. Multiply.

$$A = \ell \times w$$

$$A = \text{_____ yd} \times \text{_____ yd}$$

$$A = \text{_____ yd}^2$$

The area of the rectangle is _____ square yards.

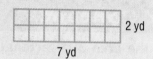

Who is Correct?

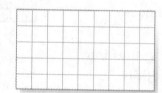
Remember that a square is a special rectangle.

Find the area of the square.

Jacqui
A = 4 + 4 = 8 ft²

Arturo
A = 4 × 4 = 16 ft²

Ryan
A = 4 × 2 = 8 ft²

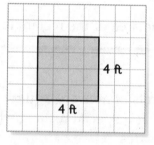
4 ft
4 ft

Circle correct answer(s). Cross out incorrect answer(s).

▶ Guided Practice

Draw a rectangle that has the given area.

1 18 square cm

2 4 cm²

3 Find the area of the rectangle.

Step 1 The length is _____ feet.

The width _____ feet.

Step 2 Substitute these values into the formula. Multiply.

$A = \ell \times w$

$A =$ _____ ft $\times$ _____ ft

$A =$ _____ ft^2

The area of the rectangle is _____ square feet.

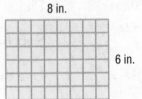

6 ft

5 ft

Find the area of each rectangle.

4 The length is _____.

The width is _____.

$A = \ell \times w$

$A =$ _____ . $\times$ _____ .

$A =$ _____

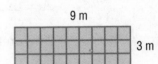

8 in.

6 in.

5 The length is _____. The width is _____.

$A =$ _____ $\times$ _____

$A =$ _____

9 m

3 m

6 $A =$ _____

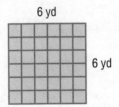

6 yd

6 yd

7 $A =$ _____

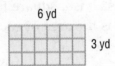

6 yd

3 yd

GO ON

8 $A =$ _____

8 mi

5 mi

9 $A =$ _____

12 km

3 km

Step (by) Step Problem-Solving Practice

Solve.

10 BASKETBALL A high-school basketball court is 94 feet long and 50 feet wide. What is the area of the court?

Problem-Solving Strategies

✓ Use a formula.

☐ Look for a pattern.

☐ Guess and check.

☐ Act it out.

☐ Solve a simpler problem.

Understand Read the problem. Write what you know.

The length of the basketball court is _____ feet.
The width of the court is _____ feet.

Plan Pick a strategy. One strategy is to use a formula.

Substitute values for length and width into the area formula.

Solve Use the formula.

$A = \ell \times w$

$A =$ _____ $\times$ _____

$A =$ _____

The area of the high-school basketball court is _____.

Check Use division or a calculator to check your multiplication.

11 CONSTRUCTION A construction crew is pouring cement for sidewalk slabs. Each slab is a square that has sides that measure 70 centimeters. What is the area of each slab?

Check off each step.

_____ Understand: I circled key words.

_____ Plan: To solve the problem, I will _____.

_____ Solve: The answer is _____.

_____ Check: I checked my multiplication with _____.

Solve.

28 cm

20 PHOTOS At the portrait studio, Ines ordered the picture of her family shown. What was the area of Ines's family portrait?

_____

36 cm

21 DOORS The screen door at Ethan's house is 32 inches wide and 85 inches tall. What is the area of Ethan's screen door?

_____

Vocabulary Check **Write the vocabulary word that completes each sentence.**

22 _____ is the number of square units needed to cover a region or a plane figure.

23 A(n) _____ is a rectangle with four equal sides.

24 A(n) _____ has opposite sides that are equal and parallel. It is a quadrilateral with four right angles.

25 Writing in Math Explain how to find the area of a rectangle.

_____

_____

_____

▶ Spiral Review

Find the area of each figure. (Lesson 9-1, p. 368)

26

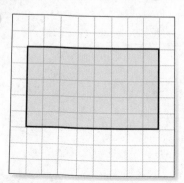

A = _____

27

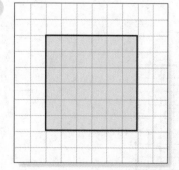

A = _____

STOP

12 ART Mrs. Brady asked her class to use an entire sheet of paper to finger paint. Each sheet of paper measured 8.5 inches by 11 inches. What was the area of the finger painting?

_____

13 Reflect Can two rectangles have the same area but different lengths and widths? Explain.

_____

_____

_____

▶ Skills, Concepts, and Problem Solving

Draw a rectangle that has the given area.

14 6 square units

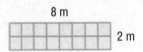

15 24 square units

Find the area of each rectangle.

16 $A =$ _____

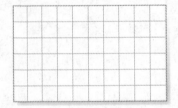
8 m
2 m

17 $A =$ _____

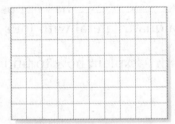
7 mi
4 mi

18 $A =$ _____

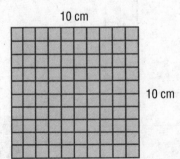
10 cm
10 cm

19 $A =$ _____

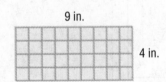
9 in.
4 in.

GO ON

Find the area of each figure.

1

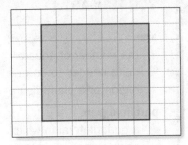

A = _____

2

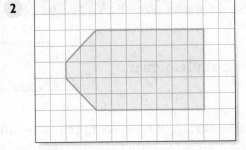

A = _____

3

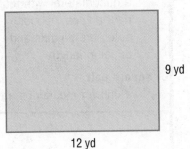

9 yd

12 yd

A = _____

4

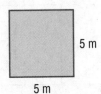

5 m

5 m

A = _____

5

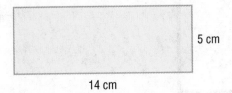

5 cm

14 cm

A = _____

6

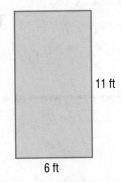

11 ft

6 ft

A = _____

Solve.

7   **DESIGN**   Lena hung a rectangular mirror on her bathroom wall.
The mirror was 85 centimeters high and 67 centimeters wide. What
was the area of Lena's mirror? _____

8   **TILES**   Lamont decorated square-shaped tiles. Each tile was
80 millimeters long and 60 millimeters wide. What was the area
of each tile? _____

Area of a Parallelogram

KEY Concept

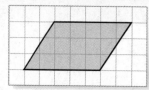

parallelogram

In a parallelogram, *b* represents the base, and *h* represents the height. Cut a triangle from the parallelogram along the dashed line. Place the triangle on the other side, next to the right edge of the parallelogram.

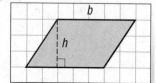 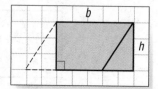

Notice that the new shape is a rectangle. So, the formulas for the areas of parallelograms and rectangles are similar.

A is the area of the parallelogram.

This is like the area of a rectangle, except the length is *b* and the width is *h*.

$$A = \ell \times w$$
$$A = b \times h$$

b is the length of the base.

h is the height.

Copyright © by The McGraw-Hill Companies, Inc.

VOCABULARY

area
the number of square units needed to cover a region or a plane figure

parallelogram
a quadrilateral in which each pair of opposite sides is parallel and equal in length

rectangle
a quadrilateral with four right angles; opposite sides are parallel and equal in length

square unit
a unit for measuring area

Example 1

Find the area of the parallelogram.

1. The base is 8 inches. The height is 7 inches.

2. Substitute these values into the formula. Multiply.

$A = b \times h$
$A = 8$ in. $\times 7$ in.
$A = 56$ in^2

The area of the parallelogram is 56 square inches.

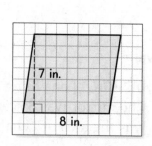

7 in.

8 in.

YOUR TURN!

Find the area of the parallelogram.

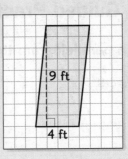

1. The base is _____ feet. The height is _____ feet.

2. Substitute these values into the formula. Multiply.

   $A = b \times h$

   $A =$ _____ ft $\times$ _____ ft

   $A =$ _____ ft^2

The area of the parallelogram is _____ square feet.

Example 2

Find the area of the parallelogram.

1. The base is 9 centimeters.
   The height is 9 centimeters.

2. Substitute these
   values into the formula.
   Multiply.

9 cm

9 cm

   $A = b \times h$
   $A = 9 \text{ cm} \times 9 \text{ cm}$
   $A = 81 \text{ cm}^2$

The area of the parallelogram is
81 square centimeters.

YOUR TURN!

Find the area of the parallelogram.

1. The base is _____ yards.
   The height is _____ yards.

7 yd

7 yd

2. Substitute these values
   into the formula. Multiply.

   $A = b \times h$

   $A =$ _____ yd $\times$ _____ yd

   $A =$ _____ yd^2

The area of the parallelogram is

_____ square yards.

Who is Correct?

Find the area of the parallelogram.

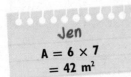

Jen

$A = 6 \times 7$
$= 42 \text{ m}^2$

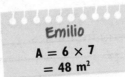

Emilio

$A = 6 \times 7$
$= 48 \text{ m}^2$

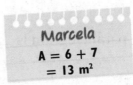

Marcela

$A = 6 + 7$
$= 13 \text{ m}^2$

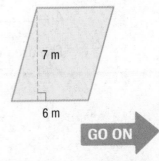

7 m

6 m

Circle correct answer(s). Cross out incorrect answer(s).

GO ON

▶ Guided Practice

Draw a parallelogram that has the given area.

1 30 cm²

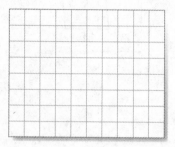

2 200 mm²

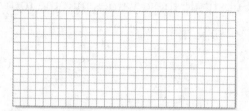

Step by Step Practice

Find the area of the parallelogram.

3

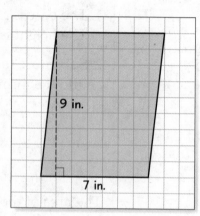

9 in.

7 in.

Step 1 The base is _____ inches.
The height is _____ inches.

Step 2 Substitute these values into the formula. Multiply.

$A = b \times h$

$A =$ _____ in. $\times$ _____ in.

$A =$ _____ in²

The area of the parallelogram is _____ square inches.

Find the area of each parallelogram.

4 The base is _____.
The height is _____.

$A = b \times h$
$A =$ _____ $\times$ _____
$A =$ _____

The area of the parallelogram is _____.

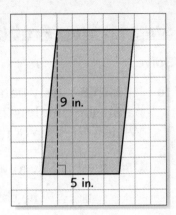
9 in.
5 in.

5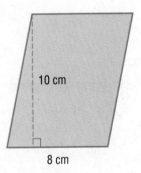
2 km
4 km

The base is _____.
The height is _____.

$A =$ _____ $\times$ _____
$A =$ _____

6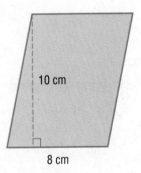
10 cm
8 cm

$A =$ _____ $\times$ _____
$A =$ _____

7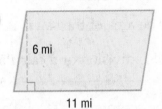
6 mi
11 mi

$A =$ _____ $\times$ _____
$A =$ _____

8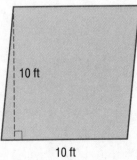
10 ft
10 ft

$A =$ _____

9
3 m
3 m

$A =$ _____

GO ON

Step by Step Problem-Solving Practice

Solve.

10 **HOBBIES** Tony bought a sail for his boat. The sail is a parallelogram. It is 21 feet wide at the base and 42 feet tall. What is the area of the new sail?

Understand Read the problem. Write what you know.

The base of the sail is _____ feet.

The height of the sail is _____ feet.

Plan Pick a strategy. One strategy is to use a formula.

Substitute the values for base and height into the area formula.

Solve Use the formula.

$A = b \times h$

$A =$ _____ × _____

$A =$ _____

The area of the sail is _____.

Check Use division or a calculator to check your multiplication.

HOBBIES A sailboat can have sails shaped like parallelograms.

11 **ART** Part of the sculpture at Jacob Park is shaped like a parallelogram. The front of this piece is 13 feet tall. It has a base of 8 feet. What is the area of the front of the sculpture? Check off each step.

_____ Understand: I circled key words.

_____ Plan: To solve the problem, I will _____.

_____ Solve: The answer is _____.

_____ Check: I checked my answer by _____.

12 **ART** Camila's class is making cardboard ornaments. Each ornament is shaped like a parallelogram with a height of 16 centimeters and a base of 9 centimeters. What is the area of each ornament?

_____

13 **Reflect** Compare the area of the parallelogram to the area of the rectangle at the right. Explain.

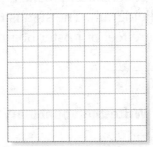

_____

_____

 Skills, Concepts, and Problem Solving

Draw a parallelogram that has the given area.

14 220 mm²

15 5 in²

Find the area of each parallelogram.

16

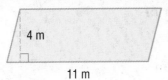
4 m
11 m

$A =$ _____

17

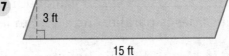
3 ft
15 ft

$A =$ _____

18

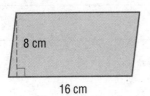
8 cm
16 cm

$A =$ _____

19

12 yd
8 yd

$A =$ _____

20

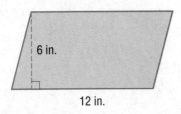
6 in.
12 in.

$A =$ _____

21

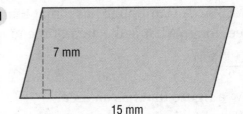
7 mm
15 mm

$A =$ _____

GO ON

Solve.

22 FARMING Mrs. Rockwell's cornfield is in the shape of a parallelogram. Refer to the photo caption at the right. What is the area of the cornfield?

_____

FARMING The height of Mrs. Rockwell's cornfield is 79 meters. The base is 52 meters.

23 HOBBIES Toby made toy wooden boats. He cuts each sail in the shape of a parallelogram. Each sail is 45 millimeters at the base and 65 millimeters tall. What is the area of each sail?

_____

24 POSTERS The cheerleaders make a Spirit Week sign in the shape of a parallelogram. The sign is 6 feet long and 4 feet tall. How many square feet of poster board do the cheerleaders use?

_____

Vocabulary Check **Write the vocabulary word that completes each sentence.**

25 A(n) _____ is a unit for measuring area.

26 A(n) _____ is a quadrilateral in which each pair of opposite sides is parallel and equal in length.

27 Writing in Math Explain how to find the area of a parallelogram with a base of 9 inches and a height of 10 inches.

_____

_____

 Spiral Review

Solve.

28 FITNESS A trampoline has a mat that is 10 feet wide and 14 feet long. What is the area of the mat? (Lesson 9-2, p. 373)

_____

29 Draw a figure that has an area of 18 square units. (Lesson 9-1, p. 368)

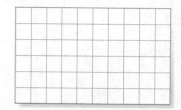

STOP

Area of a Triangle

KEY Concept

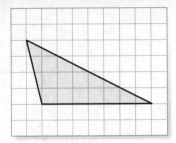

triangle

You can cut a parallelogram to create two triangles.

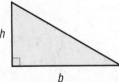 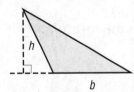

Notice you now have two triangles. Each triangle is one-half the size of the parallelogram. Remember the formula for the area of a parallelogram is $A = b \times h$

h is the height.

$$A = \frac{1}{2} \times b \times h$$

A is the area of the triangle.

b is the length of the base.

The location of the height of a triangle can vary. There are three possibilities.

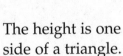 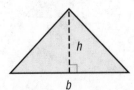

The height is one side of a triangle.

The height is inside the triangle.

The height is outside the triangle.

VOCABULARY

area
the number of square units needed to cover a region or a plane figure

parallelogram
a quadrilateral in which each pair of opposite sides is parallel and equal in length

rectangle
a quadrilateral with four right angles; opposite sides are parallel and equal in length

square unit
a unit for measuring area

triangle
a polygon with three sides and three angles

GO ON

Example 1

Find the area of the triangle.

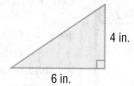

4 in.

6 in.

1. The base is 6 inches.
   The height is 4 inches.

2. Substitute these values into the formula.

   $A = \frac{1}{2} \times b \times h$

   $A = \frac{1}{2} \times 6 \text{ in.} \times 4 \text{ in.}$

3. Multiply to find the area of the triangle.

   $A = 12 \text{ in}^2$

   The area of the triangle is 12 square inches.

YOUR TURN!

Find the area of the triangle.

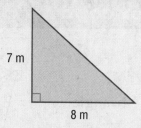

7 m

8 m

1. The base is _____ meters.

   The height is _____ meters.

2. Substitute these values into the formula.

   $A = \frac{1}{2} \times b \times h$

   $A = \frac{1}{2} \times \underline{\hspace{1cm}} \text{ m} \times \underline{\hspace{1cm}} \text{ m}$

3. Multiply to find the area of the triangle.

   $A = \underline{\hspace{1cm}} \text{ m}^2$

   The area of the triangle is

   _____ square meters.

Example 2

Find the area of the triangle.

1. The base is 3 units long.

2. The height is 4 units long.

3. Substitute these values into the formula.

   $A = \frac{1}{2} \times b \times h$

   $A = \frac{1}{2} \times 3 \times 4$

4. Multiply to find the area of the triangle.

   $A = 6 \text{ units}^2$

   The area of the triangle is 6 square units.

4

3

YOUR TURN!

Find the area of the triangle.

1. The base is _____ units long.

2. The height is _____ units long.

3. Substitute these values into the formula.

$$A = \frac{1}{2} \times b \times h$$

$$A = \frac{1}{2} \times \text{_____} \times \text{_____}$$

4. Multiply to find the area of the triangle.

$$A = \text{_____}$$

The area of the triangle is _____ square units.

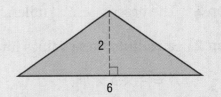

Who is Correct?

Find the area of the triangle.

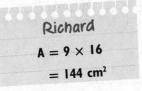

Amelia
$A = \frac{1}{2}(9 + 16)$
$= 12.5 \text{ cm}^2$

Darin
$A = \frac{1}{2} \times 9 \times 16$
$= 72 \text{ cm}^2$

Richard
$A = 9 \times 16$
$= 144 \text{ cm}^2$

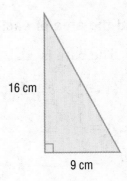

16 cm

9 cm

Circle correct answer(s). Cross out incorrect answer(s).

▶ Guided Practice

Draw a triangle that has the given area.

1 15 units²

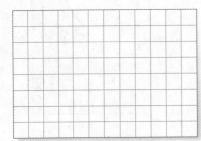

2 100 units²

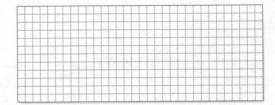

GO ON

Step by Step Practice

Find the area of the triangle.

3 **Step 1** The base is _____ miles. The height is _____ miles.

Step 2 Substitute these values into the formula.

$$A = \frac{1}{2} \times b \times h$$

$$A = \frac{1}{2} \times \text{_____} \text{ mi} \times \text{_____} \text{ mi}$$

Step 3 Multiply to find the area of the triangle.

$$A = \text{_____} \text{ mi}^2$$

The area of the triangle is _____ square miles.

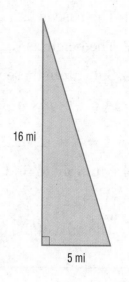

16 mi

5 mi

Find the area of each triangle.

4 The base is _____. The height is _____.

$$A = \frac{1}{2} \times b \times h$$

$$A = \frac{1}{2} \times \text{_____} \times \text{_____}$$

$$A = \text{_____}$$

The area of the triangle is _____

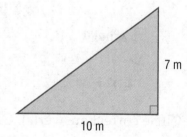

7 m

10 m

5 $A = \frac{1}{2} \times b \times h$

$$A = \frac{1}{2} \times \text{_____} \times \text{_____}$$

$$A = \text{_____}$$

The area of the triangle is _____

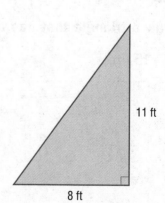

11 ft

8 ft

6 The area of the triangle is

_____

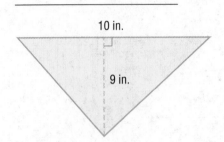

10 in.

9 in.

7 The area of the triangle is

_____

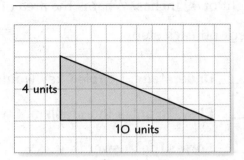

4 units

10 units

Step by Step Problem-Solving Practice

Solve.

8 **VOLUNTEERING** Lolita earned a club patch by volunteering last week. The patch is shaped like a triangle. It has a base of 53 millimeters and is 66 millimeters tall. What is the area of Lolita's patch?

Understand   Read the problem. Write what you know.

The base of the patch is _____ millimeters.
The height of the patch is _____ millimeters.

Plan   Pick a strategy. One strategy is to use a formula.

Substitute these values into the formula.

Solve   Use the formula.

$$A = \frac{1}{2} \times b \times h$$

$$A = \frac{1}{2} \times \underline{\qquad} \times \underline{\qquad}$$

$$A = \underline{\qquad}$$

The area of the patch is _____

Check   Use a calculator to check your answer.

GO ON

9 FLAGS   The Coleman Camp flag is raised every morning. The flag is in the shape of a triangle. It is 36 inches long at its base and has a height of 49 inches. What is the area of the camp flag?

Check off each step.

_____ **Understand: I circled key words.**

_____ **Plan: To solve the problem, I will** _____.

_____ **Solve: The answer is** _____.

_____ **Check: I checked my answer by** _____.

10 COSTUMES   Marita bought a triangular-shaped bandanna for her costume in the school play. The bandanna is 60 centimeters tall and has a base of 120 centimeters. What is the area of the bandanna?

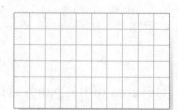

60 cm

120 cm

_____

11 Reflect   Compare the area of the triangular-shaped bandanna to the area of the parallelogram at the right.

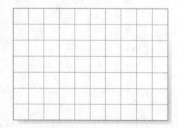

60 cm

120 cm

_____

_____

 Skills, Concepts, and Problem Solving

Draw a triangle that has the given area.

12 12 units²

13 6 units²

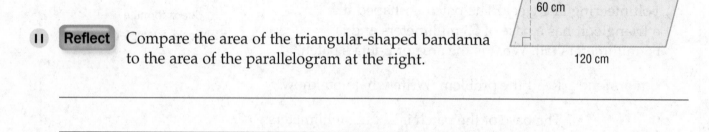

Find the area of each triangle.

14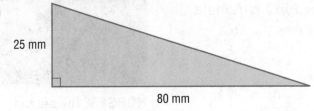

25 mm

80 mm

$A =$ _____

15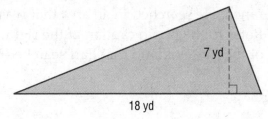

7 yd

18 yd

$A =$ _____

16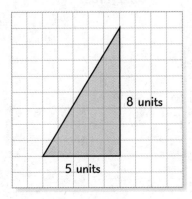

8 units

5 units

$A =$ _____

17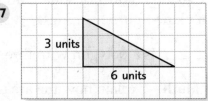

3 units

6 units

$A =$ _____

18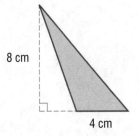

8 cm

4 cm

$A =$ _____

19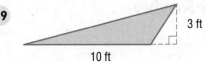

3 ft

10 ft

$A =$ _____

Solve.

20 **WOODWORKING**   For a game, Wayne made triangular-shaped wooden blocks. Each block was 48 millimeters at the base and 51 millimeters tall. What was the area of each block?

_____

GO ON

21 **HORSES**   Earl searched for a missing horse near his uncle's ranch. He searched in an area that was shaped like a triangle. Refer to the photo caption at the right. What was the area of the piece of land that Earl searched?

_____

HORSES The area in which Earl searched for the horse had a base of 886 yards and a height of 692 yards.

Vocabulary Check   **Write the vocabulary word that completes each sentence.**

22 _____ is the number of square units needed to cover a region or a plane figure.

23 A(n) _____ is a polygon with three sides and three angles.

24 **Writing in Math**   Explain how the area of a triangle is related to the area of a rectangle.

_____

_____

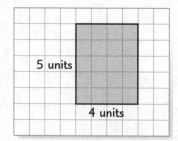 **Spiral Review**

Solve.

25 **WEATHER**   Zing turned on the weather channel. He saw a region that had a tornado watch. The region was shaped like a parallelogram. It measured 82 miles across at the base and was 56 miles long. What was the area of the region with the tornado watch? (Lesson 9-3, p. 380)

_____

26 Find the area of the rectangle. (Lessons 9-1, p. 368 and 9-2, p. 373)

5 units

4 units

The area of the rectangle is _____ square units.

Draw a figure that has the given area.

1 parallelogram, 60 square units

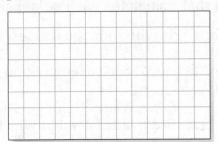

2 triangle, 15 square units

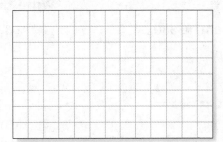

Find the area of each parallelogram.

3

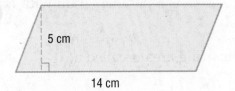

5 cm

14 cm

$A =$ _____

4

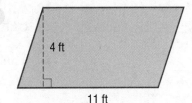

4 ft

11 ft

$A =$ _____

Find the area of each triangle.

5

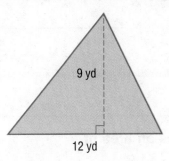

9 yd

12 yd

$A =$ _____

6

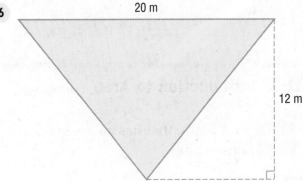

20 m

12 m

$A =$ _____

Solve.

7 **DESIGN** Ngoko hung a triangular-shaped picture in her bedroom. The picture was 60 centimeters tall and 85 centimeters at the base. What was the area of the picture? _____

8 **ARCHITECTURE** A window is in the shape of a parallelogram. The window is 3 feet tall and 4 feet at the base. What is the area of the window? _____

Vocabulary and Concept Check

area, *p. 368*

parallelogram, *p. 380*

rectangle, *p. 373*

square, *p. 373*

square unit, *p. 368*

triangle, *p. 387*

Write the vocabulary word that completes each sentence.

1. A(n) _____ is a rectangle with four equal sides.

2. The number of square units needed to cover a region or a plane figure is the _____.

3. A(n) _____ is a quadrilateral in which each pair of opposite sides is parallel and equal in length.

4. A(n) _____ is a parallelogram with four right angles.

5. The unit used for measuring area is called a(n) _____.

Identify the correct figure for each area formula.

6. $A = \ell \times w$

7. $A = \frac{1}{2} \times b \times h$

8. $A = b \times h$

_____   _____   _____

9-1 Introduction to Area (pp. 368–372)

9. Draw a figure that has an area of 23 square units.

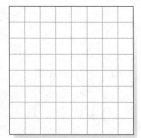

Example 1

Draw a figure that has an area of **17 square units.**

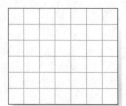

To cover an area of 17 square units, there are two options.

Cover 17 whole squares.

Cover a combination of whole and half squares that adds up to 17 squares.

Find the area of each figure.

10

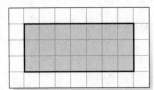

The area of the square is
_____ square units.

11

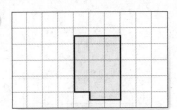

$A =$ _____

Estimate the area of each figure.

12

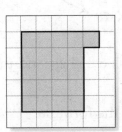

_____

_____

13

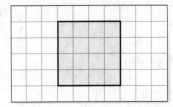

$A =$ _____

Example 2

Estimate the area of the figure.

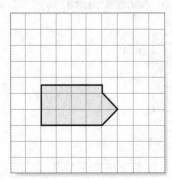

Count the number of whole squares the figure covers. The figure covers 8 whole squares.

Count the number of half squares the figure covers. The figure covers 6 half squares.

$$\frac{1}{2} + \frac{1}{2} + \frac{1}{2} + \frac{1}{2} + \frac{1}{2} + \frac{1}{2} = 3$$

Add the number of whole squares.

$8 + 3 = 11$

The area of the figure is 11 square units.

Lesson Review

9-2 Area of a Rectangle (pp. 373–378)

Find the area of each square.

14

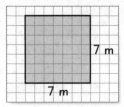

$A =$ _____

15

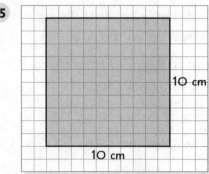

$A =$ _____

Find the area of each rectangle.

16

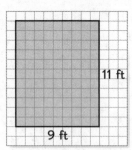

$A =$ _____

17

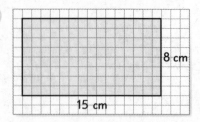

$A =$ _____

Example 3

Find the area of the square.

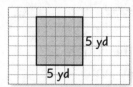

The length is 5 yards, and the width is 5 yards.

Substitute these values into the formula. Multiply.

$A = \ell \times w$
$A = 5 \text{ yd} \times 5 \text{ yd}$
$A = 25 \text{ yd}^2$

The area of the square is 25 square yards.

Example 4

Find the area of the rectangle using the formula $A = \ell \times w$.

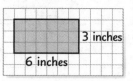

ℓ is the length of the rectangle.

w is the width of the rectangle.

A is the area of the rectangle. $\longrightarrow A = \ell \times w \longleftarrow$

The length of the rectangle is 6 inches, and the width is 3 inches.

Substitute these values into the formula. Multiply.

$A = \ell \times w$
$A = 6 \text{ in.} \times 3 \text{ in.}$
$A = 18 \text{ in}^2$

The area of the rectangle is 18 square inches.

9-3 Area of a Parallelogram (pp. 380–386)

Find the area of each parallelogram.

18

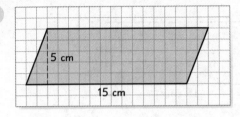

5 cm

15 cm

$A =$ _____

19

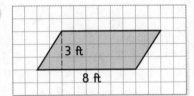

3 ft

8 ft

$A =$ _____

9-4 Area of a Triangle (pp. 387–394)

Find the area of each triangle.

20

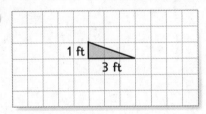

1 ft

3 ft

$A =$ _____

21

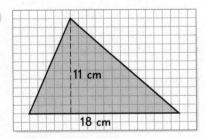

11 cm

18 cm

$A =$ _____

Example 5

What is the area of the parallelogram?

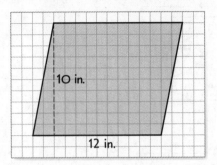

10 in.

12 in.

The base of the parallelogram is 12 inches, and the height is 10 inches.

Substitute these values into the formula. Multiply.

$A = b \times h$ The area of the
$A = 12 \text{ in.} \times 10 \text{ in.}$ parallelogram is
$A = 120 \text{ in}^2$ 120 square inches.

Example 6

What is the area of the triangle?

The base of the triangle is 10 inches, and the height of the triangle is 16 inches.

Substitute values of the base and height into the area of a triangle formula.

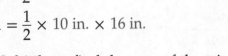

16 in.

10 in.

$A = \frac{1}{2} \times b \times h$

$A = \frac{1}{2} \times 10 \text{ in.} \times 16 \text{ in.}$

Multiply to find the area of the triangle.

$A = 80 \text{ in}^2$

The area of the triangle is 80 square inches.

Find the area of each figure.

1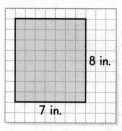

$A =$ _____

2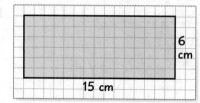

$A =$ _____

3

$A =$ _____

4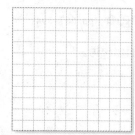

$A =$ _____

Draw a rectangle that has the given area.

5 35 units^2

6 49 units^2

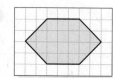

7 42 units^2

8 64 units^2

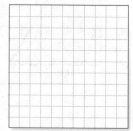

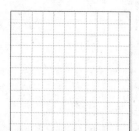

GO ON

Find the area of each parallelogram.

9

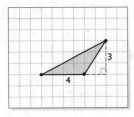

4 yd

15 yd

$A = $ _____

10

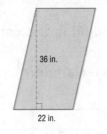

36 in.

22 in.

$A = $ _____

Find the area of each triangle.

11

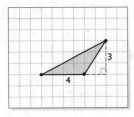

3

4

$A = $ _____

12

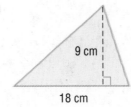

9 cm

18 cm

$A = $ _____

Solve.

13 **PETS** The floor of Jet's doghouse is a rectangle with a length of 32 inches and a width of 20 inches. What is the area of the floor of Jet's doghouse? _____

14 **BAKING** Betsy made dough cutouts in the form of triangles. Each had a base of 11 centimeters and a height of 8 centimeters. What was the area of each triangular dough cutout? _____

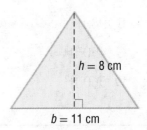

$h = 8$ cm

$b = 11$ cm

Correct the mistakes.

15 Margaret bought a square rug that measures 8 feet on each side. Later that day Margaret e-mailed her sister to describe the rug. What mistake did Margaret make?

From: Margaret
Subject: New rug
To: Sister

My new rug is 32 square feet (8 ft × 4 = 32 square feet)

Margaret

_____

_____

16 Show how Margaret should have found the area of the rug.

_____

STOP

_____

Choose the best answer and fill in the corresponding circle on the sheet at right.

1 What is the area of the rectangle?

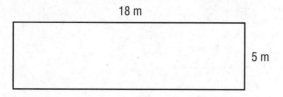

18 m

5 m

A 23 m²

B 46 m²

C 90 m²

D 100 m²

2 One wall in Mr. Viera's classroom measures 35 feet by 14 feet. What is the area of this wall?

A 49 ft²

B 98 ft²

C 490 ft²

D 560 ft²

3 Find the area of each figure. Which sentence is true?

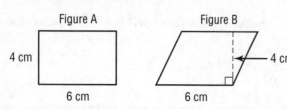

Figure A   Figure B

4 cm   4 cm

6 cm   6 cm

A Area A > Area B

B Area B > Area A

C Area A < Area B

D Area A = Area B

4 Alyssa has a parallelogram-shaped mouse pad. It has a base length of 10 inches and a height of 8 inches. What is the area of the mouse pad?

A 9 in²   **C** 36 in²

B 18 in²   **D** 80 in²

5 Find the area of each triangle in the figure.

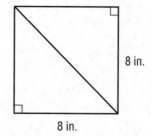

8 in.

8 in.

A 16 in²   **C** 32 in²

B 24 in²   **D** 64 in²

6 What is the area of the triangle?

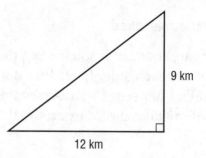

9 km

12 km

A 21 km   **C** 54 km²

B 54 km   **D** 108 km²

GO ON

7 **What is the area of the rectangle?**

3 yd

21 yd

    A 48 yards

    B 48 yd²

    C 63 yards

    D 63 yd²

8 Malik wants to make a scarf. He buys a piece of fabric that is 36 inches long and 24 inches wide. How much fabric does he have?

    A 120 inches

    B 120 square inches

    C 864 inches

    D 864 square inches

9 **What is the area of the triangle?**

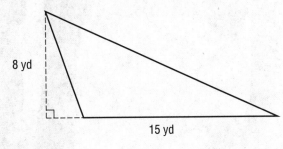

8 yd

15 yd

    A 60 yd

    B 120 yd

    C 60 yd²

    D 120 yd²

10 **What is the area of the shaded figure?**

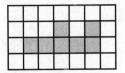

    A 7 square units

    B 8 square units

    C 22 square units

    D 28 square units

ANSWER SHEET

Directions: Fill in the circle of each correct answer.

1 Ⓐ Ⓑ Ⓒ Ⓓ
2 Ⓐ Ⓑ Ⓒ Ⓓ
3 Ⓐ Ⓑ Ⓒ Ⓓ
4 Ⓐ Ⓑ Ⓒ Ⓓ
5 Ⓐ Ⓑ Ⓒ Ⓓ
6 Ⓐ Ⓑ Ⓒ Ⓓ
7 Ⓐ Ⓑ Ⓒ Ⓓ
8 Ⓐ Ⓑ Ⓒ Ⓓ
9 Ⓐ Ⓑ Ⓒ Ⓓ
10 Ⓐ Ⓑ Ⓒ Ⓓ

Success Strategy

Try to answer every question. Work out the problem and eliminate answers you know are wrong. Do not change your answers unless you are very uncertain about your first answer choice.

STOP

Surface Area, Volume, and Measurement

How much does your backpack weigh?

How much does your dog's water dish hold? How much wrapping paper do you need to cover a box? These questions ask for measurements of weight and capacity.

STEP **1** Quiz

Are you ready for Chapter 10? Take the Online Readiness Quiz at *macmillanmh.com* to find out.

STEP **2** Preview

Get ready for Chapter 10. Review these skills and compare them with what you'll learn in this chapter.

What You Know	What You Will Learn
You know how to multiply and divide by powers of ten.	*Lesson 10-1*

What You Know

You know how to multiply and divide by powers of ten.

Examples: $4 \times 1{,}000 = 4{,}000$
$300 \div 100 = 3$

TRY IT!

1   $5 \times 100 =$ _____

2   $7 \times 10{,}000 =$ _____

3   $2{,}000 \div 100 =$ _____

4   $90{,}000 \div 10 =$ _____

You know how to multiply and divide.

Examples: $3 \times 12 = 36$
$20 \div 4 = 5$

TRY IT!

5   $4 \times 12 =$ _____

6   $36 \times 3 =$ _____

7   $64 \div 16 =$ _____

8   $72 \div 12 =$ _____

What You Will Learn

Lesson 10-1

The **metric system** is a measurement system in which units differ from the base unit by a power of ten.

1 liter of juice = 1,000 milliliters of juice

So, 4 liters of juice = $4 \times 1{,}000$, or 4,000 milliliters of juice.

Lesson 10-2

The **customary system** of measurement uses units such as a quart. You multiply or divide to change units.

4 quarts = 1 gallon
So, 20 quarts = $20 \div 4$, or 5 gallons.

Unit Conversions: Metric Capacity and Mass

KEY Concept

Prefixes used for standard units of measurement in the **metric system** always have the same meaning.

Metric prefixes indicate the place-value position of the measurement.

Prefix	Meaning in Words	Meaning in Numbers
kilo-	thousands	1,000
milli-	thousandths	0.001

The base unit of **capacity** in the metric system is the **liter**.

Metric Units for Capacity

Unit for Capacity	Abbreviation	Equivalents	Real-World Benchmark
milliliter	mL	1 mL = 0.001 L	drop of water
liter	L		sports water bottle
kiloliter	kL	1 kL = 1,000 L	bathtub filled with water

The base unit of **mass** in the metric system is the **gram**.

Metric Units for Mass

Unit for Capacity	Abbreviation	Equivalents	Real-World Benchmark
milligram	mg	1 mg = 0.001 g	grain of salt
gram	g		paper clip
kilogram	kg	1 kg = 1,000 g	digital camera

VOCABULARY

capacity
the amount of dry or liquid material a container can hold

gram
a metric unit for measuring mass

liter
a metric unit for measuring volume or capacity

mass
the amount of matter in an object

metric system
a measurement system that includes units such as meter, gram, and liter

Sometimes it is necessary to convert from one unit of measure to another. Prefixes can help you understand the relationship between two units.

Example 1

Convert 5,500 milliliters to liters.

1. Use a chart. Place 5,500 in the chart so the 0 that is farthest right is in the mL column.

2. Read the number from the chart for the conversion.

5,500 mL = 5.5 L

1,000	100	10	1	0.1	0.01	0.001
thousands	hundreds	tens	ones	tenths	hundredths	thousandths
			5 $\cdot$ 5	0		0
kilo (kL)			liter (L)			milli (mL)

YOUR TURN!

Convert 270 milliliters to liters.

1. Use a chart. Place _____ in the chart so the 0 that is farthest right is in the mL column.

2. Read the number from the chart for the conversion.

270 mL = _____ L

1,000	100	10	1	0.1	0.01	0.001
thousands	hundreds	tens	ones	tenths	hundredths	thousandths
			$\cdot$			
kilo (kL)			liter (L)			milli (mL)

Multiply to convert from larger units to smaller units.
Divide to convert from smaller units to larger units.

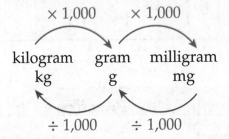

$\times 1,000 \qquad \times 1,000$

kilogram $\quad$ gram $\quad$ milligram
kg $\qquad$ g $\qquad$ mg

$\div 1,000 \qquad \div 1,000$

Example 2

Convert.

0.0027 kg = _____ g

1. To convert from kilograms to grams, multiply by 1,000.

2. Convert.

0.0027 × 1,000 = 2.7 g

YOUR TURN!

Convert.

4,600,000 mg = _____ g

1. You are converting _____ to _____. You need to _____ by 1,000.

2. Convert.

4,600,000 ÷ _____ = _____ kg

GO ON

Who is Correct?

Convert 650 liters to kiloliters.

Hugo
650 × 1,000 =
650,000

Tom
650 ÷ 10 = 65

Kwan
650 ÷ 1,000 = 0.65

Circle correct answer(s). Cross out incorrect answer(s).

 Guided Practice

Convert using a place-value chart.

1 3 kL = _____ L

1,000	100	10	1	0.1	0.01	0.001
thousands	hundreds	tens	ones	tenths	hundredths	thousandths
kilo (kL)			liter (L)			milli (mL)

2 6 mg = _____ g

1,000	100	10	1	0.1	0.01	0.001
thousands	hundreds	tens	ones	tenths	hundredths	thousandths
kilo (kg)			gram (g)			milli (mg)

 **Step by Step Practice**

Convert.

3 28 kg = _____ mg

Step 1 You are converting _____grams to _____grams.

Step 2 Since you are converting from kilograms
to milligrams, you must _____ 28 by
1,000 twice.

× 1,000   × 1,000

kilogram   gram   milligram
kg   g   mg

÷ 1,000   ÷ 1,000

Step 3 Convert.
28 × _____ × _____
28 kg = _____ mg

Convert.

4 1.2 L = _____ mL

1 L = _____ mL

1.2 _____ 1,000 = _____

1.2 L = _____ mL

5 900 mg = _____ kg

1 kg = _____ mg

900 _____ 1,000,000 = _____

900 mg = _____ kg

6 1,050 mL = _____ kL

7 0.25 g = _____ mg

8 246 mg = _____ g

9 2,010 L = _____ kL

10 936 mL = _____ L

11 880 g = _____ mg

12 404 g = _____ mg

13 31 kL = _____ L

Step by Step Problem-Solving Practice

Solve.

14 **NUTRITION** Roxanna bought a giant turkey sandwich for a party. The giant sandwich has 200,000 milligrams of protein. How many grams of protein are in the giant turkey sandwich?

Understand Read the problem. Write what you know. The giant sandwich has _____ milligrams of protein.

Plan Pick a strategy. One strategy is to use a table.

Solve Place _____ in the chart so the 0 that is farthest right is in the mL column. Read the number from the chart for the conversion. Roxanna's giant sandwich has _____ grams of protein.

Check A milligram is a smaller unit of measure than a gram, so the number of milligrams of protein should be greater than the number of grams.

Problem-Solving Strategies

☑ Use a table.

☐ Look for a pattern.

☐ Guess and check.

☐ Solve a simpler problem.

☐ Work backward.

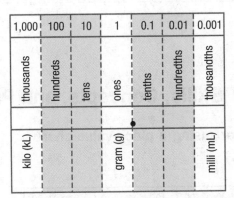

1,000	100	10	1	0.1	0.01	0.001
thousands	hundreds	tens	ones	tenths	hundredths	thousandths
			•			
kilo (kL)			gram (g)			milli (mL)

GO ON

15 **HEALTH** Jaleesa weighs 45 kilograms. How many grams does she weigh? Check off each step.

_____ Understand: I circled key words.

_____ Plan: To solve this problem, I will _____.

10,000	1,000	100	10	1	0.1	0.01	0.001
ten thousands	thousands	hundreds	tens	ones	tenths	hundredths	thousandths
	kilo (kg)			gram (g)			milli (mg)

_____ Solve: The answer is _____.

_____ Check: I checked my answer by _____.

16 **NUTRITION** Elijah drank all of the water in the bottle shown. How many milliliters of water did he drink?

_____

Bottled Water
1.2 liters

17 **Reflect** Are 65 liters equal to 0.065 kiloliters? Explain.

_____

_____

▶ Skills, Concepts, and Problem Solving

Convert using a place-value chart.

18 7 kg = _____ g

1,000	100	10	1	0.1	0.01	0.001
thousands	hundreds	tens	ones	tenths	hundredths	thousandths
kilo (kg)			gram (g)			milli (mg)

19 752 mL = _____ L

1,000	100	10	1	0.1	0.01	0.001
thousands	hundreds	tens	ones	tenths	hundredths	thousandths
kilo (kL)			liter (L)			milli (mL)

Convert.

20 0.0036 kL = _____ mL

21 1.09 g = _____ mg

22 0.01 kg = _____ g

23 15 L = _____ kL

24 0.2 mg = _____ g

25 65 kL = _____ L

26 9.4 mL = _____ kL

27 4.8 g = _____ kg

Solve.

28 **COOKING**   Anoki needed 1,500 milliliters of vegetable oil to cook a chicken for the family reunion. He bought a 2-liter bottle of oil. How many liters of oil did Anoki have left over?

_____

29 **TRAVEL**   At the airport, you can only have 32 kilograms of mass per bag. How many grams are you able to carry in each bag?

_____

30 **DRINKS**   Kendra bought a bottle of strawberry-flavored water while she waited at the airport. The bottle contained 591 milliliters of water. How many liters of water did she purchase?

_____

Vocabulary Check   **Write the vocabulary word that completes each sentence.**

31 _____ is the amount of matter in an object.

32 _____ is the amount of dry or liquid material a container can hold.

33 A(n) _____ is a metric unit for measuring volume or capacity.

34 A(n) _____ is a metric unit for measuring mass.

35 **Writing in Math**   Explain how to convert 6.07 grams to kilograms.

_____

_____

STOP

Unit Conversions: Customary Capacity and Weight

KEY Concept

The **customary system** of measurement is not based on powers of ten. It is based on numbers like 12 and 16, which have many factors.

Customary Units for Capacity

Unit for Capacity	Abbreviation	Equivalents	Real-World Benchmark
fluid ounce	fl oz		eye dropper
cup	c	1 c = 8 fl oz	coffee mug
pint	pt	1 pt = 2 c 1 pt = 16 fl oz	cereal bowl
quart	qt	1 qt = 2 pt 1 qt = 4 c 1 qt = 32 fl oz	large bottle of sports drink
gallon	gal	1 gal = 4 qt 1 gal = 8 pt 1 gal = 16 c 1 gal = 128 fl oz	milk carton

Customary Units for Weight

Unit for Capacity	Abbreviation	Equivalents	Real-World Benchmark
ounce	oz		a strawberry
pound	lb	1 lb = 16 oz	bunch of grapes
ton	T	1 T = 2,000 lb	car

VOCABULARY

benchmark
an object or number used as a guide to estimate or reference

capacity
the amount of dry or liquid material a container can hold

convert
to find an equivalent measure

customary system
a measurement system that includes units such as foot, pound, and quart

weight
a measurement that tells how heavy or light an object is

Sometimes it is necessary to **convert** from one unit of measure to another. Knowing customary conversions can help you understand the relationship between two units.

Example 1

Convert 32 pints to gallons using a table.

gallons	1	2	3	4
pints	8	16	24	32

1. 8 pints is equal to 1 gallon.

2. Fill in the table.
   1 gallon = 1 × 8 pints
   2 gallons = 2 × 8 pints
   3 gallons = 3 × 8 pints
   4 gallons = 4 × 8 pints

32 pints is equal to 4 gallons.

YOUR TURN!

Convert 3 quarts to pints using a table.

quarts	1	2	3	4
pints				

1. _____ pints is equal to 1 quart.

2. Fill in the table.

   _____ pints is equal to 3 quarts.

To convert a larger unit to a smaller unit, multiply.
To convert a smaller unit to a larger unit, divide.

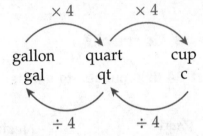

Example 2

Convert 56 fluid ounces to pints.

1. You are converting from fluid ounces to pints, which is a smaller unit to a larger unit. You need to divide.

2. 1 cup is equal to 8 fluid ounces. 1 pint is equal to 2 cups. So, 1 pint is equal to 16 fluid ounces.

   56 ÷ 16 = 3.5

So, 56 fluid ounces equals 3.5 pints.

YOUR TURN!

Convert 22 pints to gallons.

1. You are converting from _____ to _____, which is a _____ unit to a _____ unit.
   You need to _____.

2. 1 gallon is equal to _____. 1 quart is equal to _____. So, 1 gallon is equal to _____.

   22 _____ = _____

So, 22 pints equals _____.

GO ON

Example 3

Convert 2 tons to ounces.

1. You are converting from tons to ounces, which is a larger unit to a smaller unit. You need to multiply.

2. 1 ton is equal to 2,000 pounds.
   1 pound is equal to 16 ounces.

   $16 \times 2,000 = 32,000$ ounces

   So, 1 ton is equal to 32,000 ounces.

   $2 \times 32,000 = 64,000$ ounces

   So, 2 tons equals 64,000 ounces.

YOUR TURN!

Convert 3.2 tons to ounces.

1. You are converting from _____ to _____, which is a _____ unit to a _____ unit.
   You need to _____.

2. 1 ton is equal to _____.
   1 pound is equal to _____.

   3.2 _____ = _____

   So, 3.2 tons equals _____.

Who is Correct?

Convert 64 fluid ounces to quarts.

Pearl
$64 \div 16 = 4$ quarts

Nestor
$64 \div 32 = 2$ quarts

Gretchen
64 ounces $\div 8 =$
8 cups

8 cups $\div 2 =$
4 pints

4 pints $\div 2 =$
2 quarts

Circle correct answer(s). Cross out incorrect answer(s).

▶ Guided Practice

Convert using a table.

1 320 fl oz = _____ qt

quarts										
fluid ounces	32	64	96	128	160					

2 3 T = _____ oz

tons	1	2	3
ounces			

Step by Step Practice

Convert.

3 16 pt = _____ c

Step 1 You are converting from a _____ unit to a _____ unit. You need to _____.

Step 2 1 pint is equal to _____ cups.

Step 3 So, 16 pints are 16 _____ 2, or _____ cups.

Convert.

4 7 lb = _____ oz

5 2 gal = _____ qt

6 800 fl oz = _____ qt

7 10 pt = _____ c

Step by Step Problem-Solving Practice

Solve.

8 **MEASUREMENT** A bathtub for a baby can hold 7 gallons of water. How many quarts of water can the bathtub hold?

Problem-Solving Strategies

☑ Make a table.
☐ Look for a pattern.
☐ Write an equation.
☐ Solve a simpler problem.
☐ Work backward.

Understand Read the question. Write what you know.
A baby bathtub holds _____ gallons of water.

Plan Pick a strategy. One strategy is to make a table.

gallons	1	2	3	4	5	6	7
quarts	4	8	12				

Solve The table begins with the numbers 4, 8, and 12.
Continue the pattern until you find the seventh term.

The seventh term is _____. _____ quarts = _____ gal

The baby bathtub can hold _____ quarts of water.

Check A quart is a smaller unit of measurement than a gallon. So the number of quarts of water is greater than the number of gallons of water.

GO ON

9 **ZOO ANIMALS** A rhinoceros at the zoo weighs 7,000 pounds. How many tons does it weigh?

Check off each step.

_____ Understand: I circled key words.

_____ Plan: To solve this problem, I will _____.

_____ Solve: The answer is _____.

_____ Check: I checked my answer by _____.

10 **COOKING** For the baking contest this year, each baker will be given 48 ounces of flour. Galeno needs more flour than that for his recipes. He is bringing 32 ounces of flour. How many pounds of flour will Galeno have altogether? _____

11 **Reflect** Are there 64 cups in 2 gallons? Explain.

_____

_____

▶ Skills, Concepts, and Problem Solving

Convert using a table.

12 6 c = _____ fl oz

cups	1	2	3	4	5	6
fluid ounces						

13 5 lb = _____ oz

pounds	1	2	3	4	5
ounces					

Convert.

14 9,000 lb = _____ T

15 12 c = _____ qt

16 7 gal = _____ qt

17 256 oz = _____ lb

18 20 qt = _____ pt

19 8 pt = _____ fl oz

20 1.5 T = _____ oz

21 4 gal = _____ c

Solve.

22 **ART**   Lucas mixed the paint shown to make a shade of gray. How many gallons of gray paint did Lucas make?

_____

16 pints   8 pints

23 **PETS**   Vincent feeds his dog 1 cup of dog food in the morning and 1 cup of dog food in the evening. How many ounces of food will Vincent's dog eat in 14 days?

_____

Vocabulary Check   **Write the vocabulary word that completes each sentence.**

24 A(n) _____ is an object or number used as a guide for estimation or reference.

25 _____ is a measurement that tells how heavy or light an object is.

26 _____ is the amount of dry or liquid material a container can hold.

27 **Writing in Math**   Explain how to convert 12 fluid ounces to cups.

_____

_____

_____

 Spiral Review

Solve. (Lesson 10-1, p. 406)

28 **NUTRITION**   If a person consumes 71,700 grams of sugar in a year, how many kilograms of sugar was consumed? _____

Convert.

29 7 kL = _____ mL.

30 37 L = _____ kL

31 8,200 g = _____ kg

32 400 mg = _____ g

STOP

Convert.

1. 3,400 mL = _____ kL

2. 0.56 g = _____ mg

3. 332 mg = _____ g

4. 22 L = _____ mL

5. 6,050 L = _____ kL

6. 775 mL = _____ L

Convert using a table.

7. 4 gal = _____ c

gallons	1	2	3	4
cups	16			

8. 4 qt = _____ fl oz

quarts	1	2	3	4	5
fluid ounces	32				

Convert.

9. 400 lb = _____ T

10. 16 c = _____ qt

11. 5 gal = _____ qt

12. 160 oz = _____ lb

13. 15 qt = _____ pt

14. 1 pt = _____ fl oz

15. 1 T = _____ oz

16. 1 gal = _____ c

Solve.

17. **BABIES** Mora's little brother weighs 240 ounces. How many pounds does he weigh?

_____

18. **CONSTRUCTION** Edgar needs 5.3 liters of paint for his garage. How many milliliters of paint does he need?

_____

STOP

Surface Area of Rectangular Solids

KEY Concept

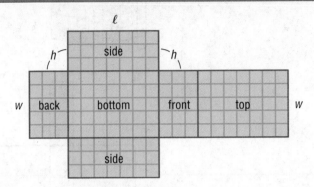

The **net** can be folded to make a rectangular prism.

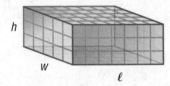

The **surface area** of a rectangular prism is the sum of the areas of all the **faces** of the figure. Surface area is measured in **square units.**

VOCABULARY

face
the flat part of a three-dimensional figure that is considered one of the sides

net
a flat pattern that can be folded to make a three-dimensional figure

square unit
a unit for measuring area

surface area
the area of the surface of a three-dimensional figure

A rectangular prism has six faces.

Example 1

What is the surface area of the rectangular prism?

1. Draw a net of the rectangular prism. Label the faces A, B, C, D, E, and F.

2. Find the area of faces A and F.

$A = \ell \times w$
$A = 2 \times 5 = 10$

3. Find the area of faces B and D.

$A = \ell \times w$
$A = 5 \times 6 = 30$

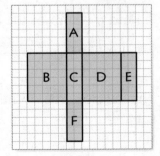

4. Find the area of faces C and E.

$A = \ell \times w$
$A = 2 \times 6 = 12$

5. Find the sum of all the areas of all the faces.

$10 + 30 + 12 + 10 + 30 + 12 = 104$

The surface area of the rectangular prism is 104 square units.

GO ON

YOUR TURN!

What is the surface area of the rectangular prism?

1. Draw a net of the rectangular prism. Label the faces A, B, C, D, E, and F.

2. Find the area of faces A and F.

$A = \ell \times w$

$A = \underline{\hspace{1cm}} \times \underline{\hspace{1cm}} = \underline{\hspace{1cm}}$

3. Find the area of faces B and D.

$A = \ell \times w$

$A = \underline{\hspace{1cm}} \times \underline{\hspace{1cm}} = \underline{\hspace{1cm}}$

4. Find the area of faces C and E.

$A = \ell \times w$

$A = \underline{\hspace{1cm}} \times \underline{\hspace{1cm}} = \underline{\hspace{1cm}}$

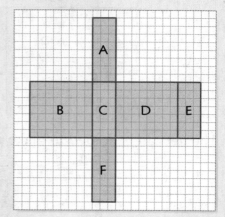

5. Find the sum of the areas of all the faces.

$\underline{\hspace{1cm}} + \underline{\hspace{1cm}} + \underline{\hspace{1cm}} + \underline{\hspace{1cm}} + \underline{\hspace{1cm}} + \underline{\hspace{1cm}} = \underline{\hspace{1cm}}$

The surface area of the rectangular prism is _____ square units.

Example 2

What is the surface area of the cube?

1. Find the area of each face.

$A = \ell \times w$
$A = 4 \times 4 = 16$

2. There are six faces on the cube. Find the sum of the areas of all six faces.

$16 + 16 + 16 + 16 + 16 + 16 = 96$
  A   B   C   D   E   F

The surface area of the cube is 96 square units.

YOUR TURN!

What is the surface area of the cube?

1. Find the area of each face.

$A = \ell \times w$
$A = \underline{\hspace{1cm}} \times \underline{\hspace{1cm}} = \underline{\hspace{1cm}}$

2. Find the sum of the areas of all six faces.

$\underline{\hspace{1cm}} + \underline{\hspace{1cm}} + \underline{\hspace{1cm}} + \underline{\hspace{1cm}} + \underline{\hspace{1cm}} + \underline{\hspace{1cm}} = \underline{\hspace{1cm}}$

The surface area of the cube is _____ square units.

Who is Correct?

What is the surface area of the rectangular prism?

Flores

A = 3 × 3 = 9 square units

A = 5 × 5 = 25 square units

A = 9 × 9 = 81 square units

A = 9 + 25 + 81
= 115 square units

Nidia

A = 3 × 9 = 21 square units

A = 3 × 5 = 15 square units

A = 5 × 9 = 45 square units

A = 21 + 15 + 45
+ 21 + 15 + 45
= 162 square units

Jermaine

A = 3 × 9 = 27 square units

A = 3 × 5 = 15 square units

A = 5 × 9 = 45 square units

A = 27 + 15 + 45
+ 27 + 15 + 45
= 174 square units

Circle correct answer(s). Cross out incorrect answer(s).

▶ Guided Practice

Draw a net for a rectangular prism with the given length, width, and height.

1. 2 × 8 × 9

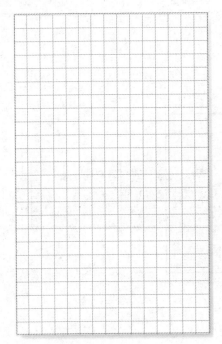

2. 5 × 4 × 7

GO ON

What is the surface area of the rectangular prism?

3

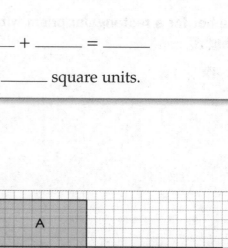

Step 1 Use a net of the rectangular prism.

Step 2 Find the area of faces A and F.

$A = \ell \times w$

$A = $ _____ $\times$ _____ $=$ _____

Step 3 Find the area of faces B and D.

$A = \ell \times w$

$A = $ _____ $\times$ _____ $=$ _____

Step 4 Find the area of faces C and E.

$A = \ell \times w$

$A = $ _____ $\times$ _____ $=$ _____

Step 5 Find the sum of the areas of all the faces.

_____ $+$ _____ $+$ _____ $+$ _____ $+$ _____ $+$ _____ $=$ _____

The surface area of the rectangular prism is _____ square units.

Find the surface area of each rectangular prism.

4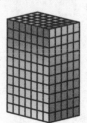

Use a net of the rectangular prism.
Follow the steps at the top of page 422
to find the surface area.

Find the area of faces A and F.

$A = \ell \times w$

$A = \underline{\hspace{1cm}} \times \underline{\hspace{1cm}} = \underline{\hspace{1cm}}$ sq. units

Find the area of faces B and D.

$A = \ell \times w$

$A = \underline{\hspace{1cm}} \times \underline{\hspace{1cm}} = \underline{\hspace{1cm}}$ sq. units

Find the area of faces C and E.

$A = \ell \times w$

$A = \underline{\hspace{1cm}} \times \underline{\hspace{1cm}} = \underline{\hspace{1cm}}$ sq. units

Find the sum of the area of all the faces.

$\underset{A}{\underline{\hspace{1cm}}} + \underset{B}{\underline{\hspace{1cm}}} + \underset{C}{\underline{\hspace{1cm}}} + \underset{D}{\underline{\hspace{1cm}}} + \underset{E}{\underline{\hspace{1cm}}} + \underset{F}{\underline{\hspace{1cm}}} = \underline{\hspace{1cm}}$ sq. units

The surface area of the rectangular prism is \underline{\hspace{1cm}} square units.

5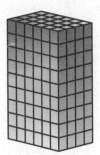

Find the area of faces A and F.

$A = \ell \times w$

$A = \underline{\hspace{1cm}} \times \underline{\hspace{1cm}} = \underline{\hspace{1cm}}$ sq. units

Find the area of faces B and D.

$A = \ell \times w$

$A = \underline{\hspace{1cm}} \times \underline{\hspace{1cm}} = \underline{\hspace{1cm}}$ sq. units

Find the area of faces C and E.

$A = \ell \times w$

$A = \underline{\hspace{1cm}} \times \underline{\hspace{1cm}} = \underline{\hspace{1cm}}$ sq. units

Find the sum of the area of all the faces.

$\underset{A}{\underline{\hspace{1cm}}} + \underset{B}{\underline{\hspace{1cm}}} + \underset{C}{\underline{\hspace{1cm}}} + \underset{D}{\underline{\hspace{1cm}}} + \underset{E}{\underline{\hspace{1cm}}} + \underset{F}{\underline{\hspace{1cm}}} = \underline{\hspace{1cm}}$ sq. units

The surface area of the rectangular prism is \underline{\hspace{1cm}} square units.

6 The surface area of the rectangular prism is \underline{\hspace{1cm}} square units.

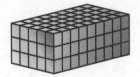

7 The surface area of the rectangular prism is \underline{\hspace{1cm}} square units.

GO ON

Solve.

8 **PRESENTS** Celia wrapped a present that was shaped like a cube. Each side measured 9 inches. What is the least amount of wrapping paper that Celia could have used to wrap the present?

Understand   Read the problem. Write what you know.
Each side of the present is _____ inches.

Plan   Pick a strategy. One strategy is to draw a diagram.
Draw a net of the cube.

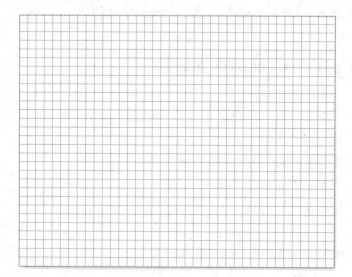

Solve   Find the area of each face.

$A = \ell \times w$
$A =$ _____ $\times$ _____ $=$ _____ sq. inches

Find the sum of the areas of all the faces.

_____ $+$ _____ $+$ _____ $+$ _____ $+$ _____ $+$
_____ $=$ _____ sq. inches

Celia used at least _____ square inches of wrapping paper.

Check   Use division and subtraction to check your multiplication and addition.

9 **DESIGN** Chase put carpet on the floor, ceiling, and all of the walls in his video room. The video room has a length of 10 feet, a width of 15 feet, and a height of 10 feet. How much carpet did Chase use for his video room? Check off each step.

_____ **Understand: I circled key words.**

_____ **Plan: To solve this problem, I will** _____.

_____ **Solve: The answer is** _____.

_____ **Check: I checked my answer by** _____.

10 **GEOMETRY** What is the surface area of a number cube that has 15 millimeter edges?

_____

11 **Reflect** Use what you know about how to find the area of triangles and rectangles to find the surface area of this triangular prism. (Hint: This triangular prism has 2 sides that are triangles and 3 sides that are rectangles.)

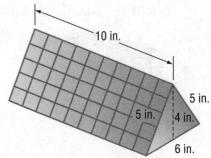

_____

_____

▶ Skills, Concepts, and Problem Solving

Draw a net for a rectangular prism with the given length, width, and height. Label the faces A, B, C, D, E, and F.

12 $3 \times 7 \times 5$

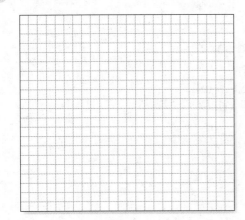

13 $4 \times 8 \times 4$

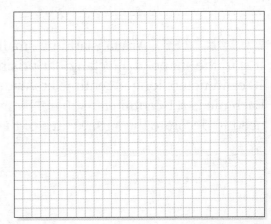

GO ON

Find the surface area of each rectangular prism.

14 The surface area of the rectangular prism is _____ square units.

15 The surface area of the rectangular prism is _____ square units.

16 The surface area of the rectangular prism is _____ square units.

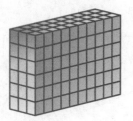

17 The surface area of the rectangular prism is _____ square units.

Solve.

18 **JEWELRY** Daisy made the jewelry box shown. The sides measure 18 centimeters each. What is the surface area of the jewelry box?

_____

JEWELRY Daisy's jewelry box is shaped like a cube.

19 **ART** Quinn decorated a rectangular-shaped chest with wallpaper. The length of the chest is 3 feet, the width is 5 feet, and the height is 2 feet. What is the least amount of wallpaper Quinn used?

_____

20 **MOSAIC TILES** Arleta is adding small tiles to a soap dish. It is in the shape of a rectangular prism. The soap dish is 10 centimeters long, 8 centimeters wide, and 3 centimeters tall. Each tile is a 1-centimeter square. How many tiles will Arleta need to decorate the surface area of the soap dish?

_____

Vocabulary Check   **Write the vocabulary word that completes each sentence.**

21 A(n) _____ is a flat pattern that can be folded to make a three-dimensional figure.

22 A(n) _____ is the flat side of a three-dimensional figure that is considered one of the sides.

23 _____ is the area of the surface of a three-dimensional figure.

24 **Writing in Math**   Explain how to find the surface area of a rectangular prism.

_____

_____

 Spiral Review

Convert using a place value chart. (Lesson 10-1, p. 406)

25 0.49 L = _____ mL

1,000	100	10	1	0.1	0.01	0.001
thousands	hundreds	tens	ones	tenths	hundredths	thousandths
kilo (kL)			liter (L)			milli (mL)

26 78,000 mg = _____ kg

1,000	100	10	1	0.1	0.01	0.001
thousands	hundreds	tens	ones	tenths	hundredths	thousandths
kilo (kg)			gram (g)			milli (mg)

Convert. (Lesson 10-2, p. 412)

27 3,000 lb = _____ T

28 16 pt = _____ gal

29 6 c = _____ fl oz

30 8 pt = _____ c

31 **PACKAGING**   Sancho bought a 2-pound box of chocolates. How many ounces of chocolate did he buy? _____

STOP

Introduction to Volume

KEY Concept

The amount of space inside a three-dimensional figure is its **volume**. Volume is measured in **cubic units**. To find the volume of a solid figure, determine the number of cubic units the solid figure contains.

One way to determine the volume of a **rectangular prism** is to think about the number of **cubes** in each layer.

This figure has 2 layers. Each layer has 10 cubes.

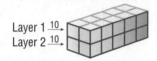

2 layers of 10 cubes = 10 + 10 = 20

This rectangular prism has a volume of 20 cubic units.

VOCABULARY

cube
a rectangular prism with six congruent square faces

cubic unit
a unit for measuring volume

rectangular prism
a three-dimensional figure with six faces that are rectangles

volume
the number of cubic units needed to fill a three-dimensional figure or solid figure

The volume of a figure is related to its dimensions, or length, width, and height.

Example 1

Find the volume of the rectangular prism.

1. Count the number of cube layers in the prism. There are 2 layers of cubes.

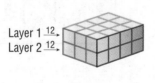

2. Count the number of cubes in the top layer. There are 12 cubes in the top layer.

3. Each layer has the same number of cubes. There are 12 + 12 = 24 cubes.

The volume of the rectangular prism is 24 cubic units.

YOUR TURN!

Find the volume of the rectangular prism.

1. How many layers of cubes are in the prism?

   _____

2. How many cubes are in the top layer?

   _____

3. There are _____ + _____ + _____ = _____ cubes.

The volume of the rectangular prism is _____ cubic units.

Example 2

Find the volume of the rectangular prism.

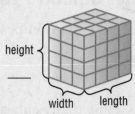

1. Look at the top layer of cubes on the prism. Finding the area of the rectangle would tell you how many cubes are on that layer.

2. Area = $\ell \cdot w$, so the area of the top layer is $4 \times 5 = 20$. There are 20 cubes on the top layer.

3. Each layer has the same number of cubes. There are 2 layers, so there are $20 + 20 = 40$ cubes in the prism.

4. The volume of the rectangular prism is 40 cubic units.

YOUR TURN!

Find the volume of the rectangular prism.

____

1. Look at the top layer of cubes on the prism. The length of the prism has _____ cubes. The width of the prism has _____ cubes.

2. Area = $\ell \cdot w$, so the area of the top layer is _____ × _____ = _____. There are _____ cubes on the top layer.

3. Each layer has the same number of cubes. There are _____ layers, so there are _____ + _____ + _____ + _____ = _____ cubes in the prism.

4. The volume of the rectangular prism is _____ cubic units.

Who is Correct?

Find the volume of the rectangular prism.

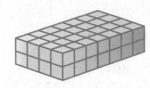

Malina

Each layer has 28 cubes. There are 2 layers. The volume is 56 cubic units.

Simon

The length has 4 cubes. The width has 7 cubes. The height has 2 cubes.

$4 \times 7 \times 2 = 56$. The volume is 56 cubic units.

Erin

The length has 4 cubes. The width has 7 cubes. The height has 2 cubes.

$4 + 7 + 2 = 13$. The volume is 13 cubic units.

Circle correct answer(s). Cross out incorrect answer(s).

GO ON

▶ Guided Practice

1 How many cubes are in this rectangular prism?

_____

2 How many cubes are in this rectangular prism?

_____

> Remember, you can find the volume of a solid figure by counting the number of cubic units it contains.

Step (by) Step Practice

3 Find the volume of the rectangular prism.

Step 1 Count the number of cubes along the length.

The length of the rectangular prism has _____ cubes.

Step 2 Count the number of cubes along the width.

The width of the rectangular prism has _____ cubes.

Step 3 The area of the top layer is _____ × _____ = _____.

Step 4 There are _____ layers in the prism.

_____ + _____ + _____ = _____ cubes in the prism.

The volume of the rectangular prism is _____ cubic units.

Find the volume of each rectangular prism.

4 Count the number of cubes along the length, width, and height of the rectangular prism.

Find the area of the top layer. Add that number four times. The volume of the rectangular prism is _____ cubic units.

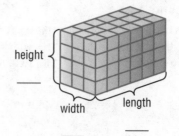

height

_____

width

length

_____   _____

5 Count the number of cubes along the length, width, and height of the rectangular prism.

Find the area of the top layer. Add that number eight times. The volume of the rectangular prism is _____ cubic units.

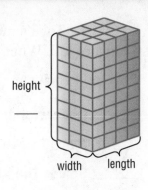

height {

width   length

_____   _____

6 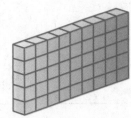 The volume of the rectangular prism is _____ cubic units.

7  The volume of the rectangular prism is _____ cubic units.

Step by Step Problem-Solving Practice

Solve.

8 **GIFTS** A gift box that is shaped like a rectangular prism has a length of 5 feet, a width of 2 feet, and a height of 8 feet. What is its volume?

Problem-Solving Strategies
- ☐ Draw a diagram.
- ☐ Look for a pattern.
- ☑ Use a model.
- ☐ Solve a simpler problem.
- ☐ Work backward.

Understand Read the problem. Write what you know.

A gift box has a length of _____ feet,

a width of _____ feet, and a height of _____ feet.

Plan Pick a strategy. One strategy is to use a model.

Solve Use unit blocks to build the rectangular prism. Count the number of blocks used.

Check The length is _____ feet.

The width is _____ feet.

There are _____ layers.

Multiply then add.

_____ × _____ = _____

_____ + _____ + _____ + _____ + _____ + _____ + _____ + _____ = _____

The volume of the gift box is _____ cubic feet.

GO ON

9 MODELS Toya's model house is 15 inches long, 20 inches wide, and 10 inches tall. What is the volume of Toya's house?

Check off each step.

_____ Understand: I circled key words.

_____ Plan: To solve this problem, I will _____.

_____ Solve: The answer is _____.

_____ Check: I checked my answer by _____.

10 PETS Bret's doghouse is 2 meters long, 2 meters wide, and 1 meter tall. What is the volume of Bret's doghouse?

_____

11 Reflect Give the length, width, and height of a rectangular prism that has a volume of 36 cubic units. Explain.

_____

_____

 Skills, Concepts, and Problem Solving

Find the volume of each rectangular prism.

12

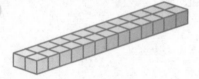

The volume of the rectangular prism is _____ cubic units.

13

The volume of the rectangular prism is _____ cubic units.

14 The volume of the rectangular prism is _____ cubic units.

15

The volume of the rectangular prism is _____ cubic units.

Solve.

16 **PACKAGING**   What is the volume of the package shown at the right?

_____

17 **ART**   A box of art tools is 8 centimeters long, 4 centimeters wide, and 5 centimeters tall. What is the volume of the art box?

_____

PACKAGING The package is 10 inches long, 7 inches wide, and 3 inches tall.

Vocabulary Check   **Write the vocabulary word that completes each sentence.**

18 A(n) _____ is a unit for measuring volume.

19 _____ is the number of cubic units needed to fill a three-dimensional figure or solid figure.

20 **Writing in Math**   Explain how to find the volume of a rectangular prism.

_____

▶ **Spiral Review**

Draw a net for a rectangular prism with the given length, width, and height. Label the faces A, B, C, D, E, and F. (Lesson 10-3, p. 419)

21 $3 \times 5 \times 6$

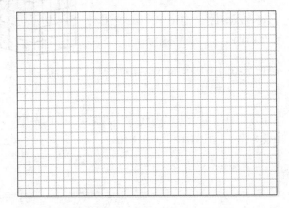

22 $2 \times 3 \times 1$

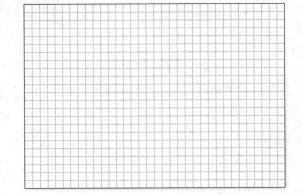

STOP

Volume of Rectangular Solids

KEY Concept

The amount of space inside a three-dimensional figure is the **volume** of the figure.

The volume of a rectangular solid is the product of its length, width, and height.

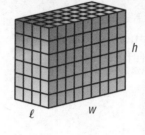

V is the volume of the solid figure.

w is the width.

$$V = \ell \times w \times h \quad \text{or} \quad V = \ell wh$$

ℓ is the length. *h* is the height.

VOCABULARY

cube
a three-dimensional figure with six congruent square faces

cubic unit
a unit for measuring volume

volume
the number of cubic units needed to fill a three-dimensional figure or solid figure

Volume is measured in **cubic units.**

Example 1

What is the volume of the rectangular prism?

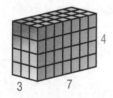

1. The length of the cube is 3 units.

    The width of the cube is 7 units.

    The height of the cube is 4 units.

2. Substitute the length, width, and height into the volume formula.

    $V = \ell \times w \times h$
    $V = 3 \times 7 \times 4$

3. Multiply.

    $V = 84$

The volume of the rectangular prism is 84 cubic units.

YOUR TURN!

What is the volume of the rectangular prism?

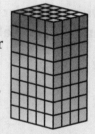

1. The length of the rectangular prism is _____ units.

    The width of the rectangular prism is _____ units.

    The height of the rectangular prism is _____ units.

2. Substitute the length, width, and height into the volume formula.

    $V = \ell \times w \times h$
    $V = \underline{\quad} \times \underline{\quad} \times \underline{\quad}$

3. Multiply.

    $V = \underline{\quad}$

The volume of the rectangular prism is _____ cubic units.

Example 2

What is the volume of the cube?

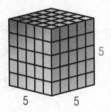

1. The length of the cube is 5 units.

   The width of the cube is 5 units.

   The height of the cube is 5 units.

2. Substitute the length, width, and height into the volume formula.

   $V = \ell \times w \times h$
   $V = 5 \times 5 \times 5$
   $V = 5^3$

3. Multiply.

   $V = 125$

The volume of the cube is 125 cubic units.

YOUR TURN!

What is the volume of the cube?

1. The length of the cube is _____ units.

   The width of the cube is _____ units.

   The height of the cube is _____ units.

2. Substitute the length, width, and height into the volume formula.

   $V = \ell \times w \times h$
   $V = \_\_\_\_ \times \_\_\_\_ \times \_\_\_\_$
   $V = \_\_\_\_$

3. Multiply.

   $V = \_\_\_\_$

The volume of the cube is _____ cubic units.

Who is Correct?

What is the volume of the rectangular prism?

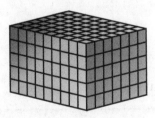

Ramiro

V = 7 × 9 × 5
= 315 cubic units

Kristen

V = 63 + 45 + 35
+ 63 + 45 + 35
= 286 cubic units

Demitri

V = 7 × 9 × 5
= 385 cubic units

Circle correct answer(s). Cross out incorrect answer(s).

GO ON

▶ Guided Practice

1 How many cubes are in this rectangular prism? _____

Check your answer. Remember, you can find the volume of a solid figure by counting the number of cubic units it contains.

2 How many cubes are in this rectangular prism? _____

Step by Step Practice

Find the volume of the rectangular prism.

3

Step 1 The length of the rectangular prism is _____ units.

The width of the rectangular prism is _____ units.

The height of the rectangular prism is _____ units.

Step 2 Substitute the length, width, and height into the volume formula.

$V = \ell \times w \times h$

$V = \underline{\quad} \times \underline{\quad} \times \underline{\quad}$

Step 3 Multiply.

$V = \underline{\quad}$

The volume of the rectangular prism is _____ cubic units.

Find the volume of each rectangular prism.

4 Substitute the length, width, and height into the volume formula. Then multiply.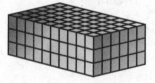

$V = \ell \times w \times h$

$V = \underline{\quad} \times \underline{\quad} \times \underline{\quad}$

$V = \underline{\quad}$

The volume of the rectangular prism is _____ cubic units.

5 $V = \ell \times w \times h$

$V = \underline{\hphantom{XXX}} \times \underline{\hphantom{XXX}} \times \underline{\hphantom{XXX}}$

$V = \underline{\hphantom{XXX}}$

The volume of the rectangular prism is _____ cubic units.

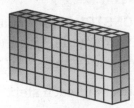

6 The volume of the rectangular prism is _____ cubic units.

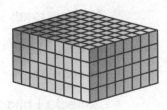

7 The volume of the rectangular prism is _____ cubic units.

Step by Step Problem-Solving Practice

Solve.

8 **HOUSES** Hillary has an air humidifier in her bedroom that is 10 inches long, 17 inches wide, and 10 inches tall. What is the volume of Hillary's air humidifier?

Understand Read the problem. Write what you know.

The air humidifier has a length of _____ inches, a width of _____ inches, and a height of _____ inches.

Plan Pick a strategy. One strategy is to use a model.

Stack cubes to model the air humidifier.

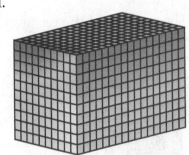

Solve Use the formula.

$V = \ell \times w \times h$

$V = \underline{\hphantom{XXX}}$ in. $\times \underline{\hphantom{XXX}}$ in. $\times \underline{\hphantom{XXX}}$ in.

$V = \underline{\hphantom{XXXX}}$ cubic inches

The volume of Hillary's air humidifier is _____ cubic inches.

Check Use division to check your multiplication.

Problem-Solving Strategies
☑ Use a model.
☐ Look for a pattern.
☐ Guess and check.
☐ Act it out.
☐ Work backward.

GO ON

9 **CONSTRUCTION** Sanouk's family has a storage shed that is 5 yards wide, 28 yards long, and 5 yards high. What is the volume of the shed?

Check off each step.

_____ **Understand: I circled key words.**

_____ **Plan: To solve this problem, I will** _____.

_____ **Solve: The answer is** _____.

_____ **Check: I checked my answer by** _____.

10 **PACKAGING** Mrs. Reynolds put together a card box for her daughter's graduation party. The card box was shaped like a cube. Each side measured 50 centimeters. What was the volume of the card box? _____

11 **Reflect** Compare the volume of the rectangular prism shown at the right to its surface area.

_____

_____

▶ Skills, Concepts, and Problem Solving

Find the volume of each rectangular prism.

12 The volume of the rectangular prism is _____ cubic units.

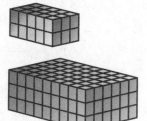

13 The volume of the rectangular prism is _____ cubic units.

14 The volume of the rectangular prism is _____ cubic units.

15 The volume of the rectangular prism is _____ cubic units.

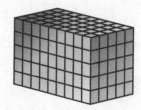

Solve.

16 **CONSTRUCTION**   Marco's father built the tree house shown at the right. What was the volume of the tree house?

_____

17 **COLLECTIONS**   Brittany used a shoe box for her rock collection. The shoe box was 12 inches long, 6 inches wide, and 5 inches high. What was the volume of the shoe box?

_____

CONSTRUCTION The tree house that Marco's father built had a height of 7 feet, a length of 10 feet, and a width of 6 feet.

Vocabulary Check   **Write the vocabulary word that completes each sentence.**

18 _____ is a unit for measuring volume.

19 _____ is the number of cubic units needed to fill a three-dimensional figure or solid figure.

20 **Writing in Math**   Explain how to find the volume of a rectangular prism.

_____

_____

_____

▶ Spiral Review

Solve. (Lesson 10-3, p. 419)

21 **MUSIC**   Alvar's stereo speakers are cube shaped. Each side measures 20 millimeters. What is the surface area of each speaker?

_____

22 Draw a net for a rectangular prism with a length of 10, a width of 7, and a height of 4.

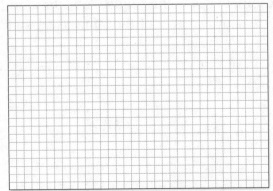

STOP

Draw a net for a rectangular prism with the given length, width, and height.

1 $3 \times 7 \times 6$

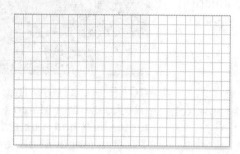

2 $2 \times 8 \times 5$

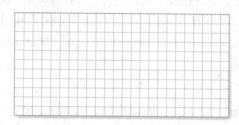

Find the surface area and the volume of each rectangular prism.

3 The surface area of the rectangular prism is _____ square units.

4 The volume of the rectangular prism is _____ cubic units.

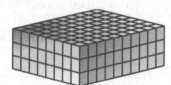

5 The surface area of the rectangular prism is _____ square units.

6 The volume of the rectangular prism is _____ cubic units.

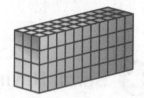

Solve.

7 **GIFTS** Melinda is wrapping a present for her sister's graduation. The gift is a book that is 6 inches long, 5 inches wide, and 2 inches high. What is the surface area of the book?

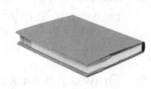

_____

8 **STORAGE** A storage cabinet is 24 inches wide, 26 inches long, and 40 inches high. What is the surface area of the storage cabinet?

_____

Vocabulary and Concept Check

benchmark, *p. 412*

capacity, *p. 406*

cube, *p. 428*

cubic unit, *p. 428*

customary system, *p. 412*

face, *p. 419*

gram, *p. 406*

liter, *p. 406*

mass, *p. 406*

metric system, *p. 406*

net, *p. 419*

rectangular prism, *p. 428*

square unit, *p. 419*

surface area, *p. 419*

volume, *p. 428*

weight, *p. 412*

Write the vocabulary word that completes each sentence.

1 A _____ is the metric unit for measuring mass.

2 _____ is the area of the surface of a three-dimensional figure.

3 Volume is measured in _____ .

4 The _____ is a measurement system that includes units such as grams and liters.

5 The _____ is a measurement system that includes units such as gallons and pounds.

6 A _____ is a flat pattern that can be folded to make a three-dimensional figure.

Label each diagram below. Write the correct vocabulary term in each blank.

7 The net shown is of a _____ that has 6 _____ .

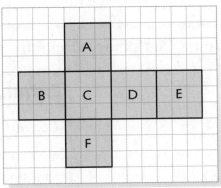

8 _____

9 _____

$$1 \text{ g} = 1,000 \text{ mg}$$

10 _____

11 _____

$$1 \text{ kL} = 1,000 \text{ L}$$

Lesson Review

10-1 Unit Conversions: Metric Capacity and Mass (pp. 406–411)

Convert using a place-value chart.

12 9 kg = _____ g

1,000	100	10	1	0.1	0.01	0.001
thousands	hundreds	tens	ones	tenths	hundredths	thousandths
kilo (kg)			gram (g) •			milli (mg)

13 30 L = _____ mL

14 2 g = _____ kg

15 15 kg = _____ g

16 45 g = _____ kg

17 12,400 mL = _____ L

Example 1

Convert 2,503 milliliters to liters.

Use a chart. Place 2,503 in the chart so the 3 that is farthest right is in the mL column.

1,000	100	10	1	0.1	0.01	0.001
thousands	hundreds	tens	ones	tenths	hundredths	thousandths
kilo (kL)			2 • liters (L)	5	0	3 milli (mL)

Read the number from the chart for the conversion.

2,503 mL = 2.503 L

10-2 Unit Conversions: Customary Capacity and Weight (pp. 412–417)

Convert.

18 20 qt = _____ gal

19 3 T = _____ lb

20 2 c = _____ fl oz

21 16,000 lb = _____ T

22 16 c = _____ qt

23 224 oz = _____ lb

Example 2

Convert 5 pounds to ounces.

You are converting from a larger unit to a smaller unit. You need to multiply. There are 16 ounces in 1 pound.

$5 \times 16 = 80$

5 lb = 80 oz

10-3 Surface Area of Rectangular Solids (pp. 419–427)

Find the surface area of each rectangular prism.

24

The surface area of the cube is _____ square units.

25

The surface area of the cube is _____ square units.

Example 3

What is the surface area of the rectangular prism?

Draw a net of the rectangular prism. Label the faces A, B, C, D, E, and F.

Find the area of faces A and F.

$A = \ell \times w$
$A = 2 \times 4 = 8$

Find the area of faces B and D.

$A = \ell \times w$
$A = 4 \times 5 = 20$

Find the area of faces C and E.

$A = \ell \times w$
$A = 2 \times 5 = 10$

Find the sum of the areas of all the faces.

8 + 20 + 10 + 8 + 20 + 10 = 76

The surface area of the rectangular prism is 76 square units.

Find the surface area of each rectangular prism.

26

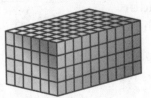

The surface area of the rectangular prism is _____ square units.

27

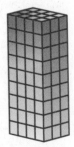

The surface area of the rectangular prism is _____ square units.

10-4 Introduction to Volume

(pp. 428–433)

28 Find the volume of the rectangular prism.

_____

_____

29 Find the volume of the rectangular prism.

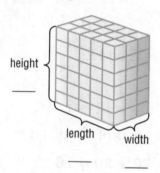

height

length   width

_____ _____ _____

_____

_____

30 Find the volume of the rectangular prism.

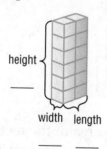

height

width   length

_____ _____

_____

_____

Example 4

Find the volume of the rectangular prism.

Count the number of layers of cubes in the prism. There are 5 layers of cubes in the rectangular prism.

Count the number of cubes in the top layer. There are 6 cubes in the top layer.

Each layer has the same number of cubes. There are 6 + 6 + 6 + 6 + 6 cubes.

The volume of the rectangular prism is 30 cubic units.

Example 5

Find the volume of the rectangular prism.

Count the number of cubes along the length. The length of the rectangular prism has 3 cubes.

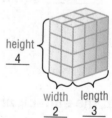

height
4

width   length
2   3

Count the number of cubes along the width. The width of the rectangular prism has 2 cubes.

Count the number of cubes along the height. The height of the rectangular prism has 4 cubes.

Find the area of the top layer.

3 × 2 = 6

Add that number 4 times.

6 + 6 + 6 + 6 = 24

The volume of the rectangular prism is 24 cubic units.

10-5 Volume of Rectangular Solids (pp. 434–439)

Find the volume of each rectangular prism.

31

The volume of the rectangular prism is _____ cubic units.

32

The volume of the rectangular prism is _____ cubic centimeters.

33

The volume of the rectangular prism is _____ cubic units.

34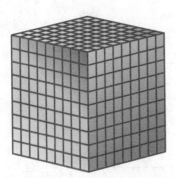

The volume of the rectangular prism is _____ cubic units.

Example 6

Find the volume of the rectangular solid using the formula below:

ℓ is the length. h is the height.

$$V = \ell \times w \times h$$

V is the volume of the solid figure. w is the width.

The length of the cube is 4 units.

The width of the cube is 9 units.

The height of the cube is 2 units.

Substitute the length, width, and height into the volume formula.

$V = \ell \times w \times h$
$V = 4 \times 9 \times 2$
$V = 72$

The volume of the rectangular prism is 72 cubic units.

Convert using a place-value chart or table.

1 6,300 mg = _____ g

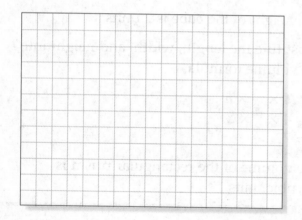

2 7 gal = _____ qt

gallons	1	2	3	4	5	6	7
quarts							

Draw a net for a rectangular prism with the given length, width, and height.

3 $2 \times 4 \times 5$

4 $3 \times 3 \times 3$

Convert.

5 400 g = _____ kg

6 2.2 L = _____ mL

7 15 kL = _____ L

8 11.2 g = _____ mg

9 250 mL = _____ L

10 0.049 kg = _____ g

11 5 T = _____ oz

12 10 gal = _____ fl oz

13 16 pt = _____ qt

14 5 lb = _____ oz

15 6,000 lb = _____ T

16 8 gal = _____ c

GO ON

Find the surface area of each rectangular prism.

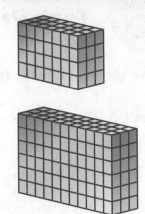

17 The surface area of the rectangular prism is
_____ square units.

18 The surface area of the rectangular prism is
_____ square units.

Find the volume of each rectangular prism.

19 The volume of the rectangular
prism is _____ cubic units.

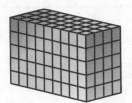

20 The volume of the rectangular
prism is _____ cubic units.

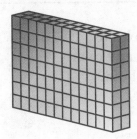

Solve.

21 **PETS**   Jet's doghouse is a rectangular prism. It has a
length of 30 inches, a width of 20 inches, and a height
of 30 inches. What is the volume of Jet's doghouse?

_____

22 **SEWING**   Mrs. Larson is making a cloth cover for her
daughter's storage trunk. The storage trunk is 4 feet
long, 2 feet wide, and 3 feet high. How much cloth will
Mrs. Larson need to cover the surface area of the trunk?

_____

Correct the mistake.

23 Tyrell bought a 2-gallon container of laundry detergent.
He told his sister that his purchase was equal to four 1-quart
containers of laundry detergent. What mistake did Tyrell make?

_____

Show how to find how many quarts are equal to 2 gallons.

_____

Choose the best answer and fill in the corresponding circle on the sheet at the right.

1 Which has a mass of about 1 kilogram?

    A a grain of salt

    B a digital camera

    C a small paper clip

    D a granola bar

2 The pitcher can hold 16 cups. How many quarts can the pitcher hold?

    A 2

    B 4

    C 6

    D 8

3 Which symbol makes this sentence true?

    2 pounds ☐ 40 ounces

    A <   **C** +

    B >   **D** =

4 Nathan sells lemonade in glasses that hold 12 fluid ounces. How many glasses can be filled with 3 gallons of lemonade?

    A 22 glasses   **C** 28 glasses

    B 32 glasses   **D** 34 glasses

5 What is the volume of the solid figure?

    A 9 cubic units   **C** 16 cubic units

    B 12 cubic units   **D** 18 cubic units

6 What is the volume of the solid figure?

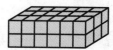

    A 12 cubic units   **C** 30 cubic units

    B 18 cubic units   **D** 36 cubic units

7 8,039 mL = _____ L

    A 803.9   **C** 8.039

    B 80.39   **D** 0.008039

8 A baby weighs 4 kilograms. How much does the baby weigh in grams?

    A 1,200 grams

    B 28 grams

    C 400 grams

    D 4,000 grams

GO ON

9 Silvia has a closed shoe box that measures 10 inches by 7 inches by 5 inches. What is the volume of the shoe box?

    A   22 cubic inches

    B   350 cubic inches

    C   350 square inches

    D   350 inches

---

10 What is the surface area of the rectangular solid?

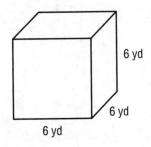

6 yd

6 yd

6 yd

    A   18 square yards

    B   186 square yards

    C   216 square yards

    D   248 square yards

---

11 Which symbol makes this sentence true?

2 liters ☐ 200 milliliters

    A   <   C   −

    B   >   D   =

12 4 pounds = _____ ounces

    A   2   C   16

    B   8   D   64

ANSWER SHEET

Directions: Fill in the circle of each correct answer.

1   Ⓐ Ⓑ Ⓒ Ⓓ
2   Ⓐ Ⓑ Ⓒ Ⓓ
3   Ⓐ Ⓑ Ⓒ Ⓓ
4   Ⓐ Ⓑ Ⓒ Ⓓ
5   Ⓐ Ⓑ Ⓒ Ⓓ
6   Ⓐ Ⓑ Ⓒ Ⓓ
7   Ⓐ Ⓑ Ⓒ Ⓓ
8   Ⓐ Ⓑ Ⓒ Ⓓ
9   Ⓐ Ⓑ Ⓒ Ⓓ
10   Ⓐ Ⓑ Ⓒ Ⓓ
11   Ⓐ Ⓑ Ⓒ Ⓓ
12   Ⓐ Ⓑ Ⓒ Ⓓ

Success Strategy

Double check your answers after you finish. Read each problem and all of the answer choices. Put your finger on each bubble you filled in to make sure it matches the answer for each problem.

STOP

Index